M.BROCK FENTON

Just Bats

UNIVERSITY OF TORONTO PRESS

TORONTO BUFFALO LONDON

©University of Toronto Press 1983
Toronto Buffalo London
Printed in Canada
Reprinted 1984, 1988, 1991

ISBN 0-8020-2452-1 cloth
ISBN 0-8020-6464-7 paper

Canadian Cataloguing in Publication Data

Fenton, M.B. (Melville Brockett), 1943-
    Just bats

    Includes index.
    ISBN 0-8020-2452-1 (bound). - ISBN 0-8020-6464-7 (pbk.)

    1. Bats. I. Title.

QL737.C5F45   599.4   C82-095126-9

This book has been published with
the assistance of the Canada Council
and the Ontario Arts Council
under their block grant programs.

(Frontispiece) This Gambian Epauletted Fruit Bat (100 grams) from Zimbabwe
wraps itself in its wings when roosting during the day. Many but not all bats hang
upside-down, a posture presumably suited to their modified forelimbs. The claw on
the second finger indicates that the bat is a pteropodid.

Contents

Why Bats?

The reasons for my interest in bats are many, but conveying them to the non-enthusiast is not so easy. As a child I was fascinated by living organisms, including bats, and I have vivid memories of being bitten by a bat that was roosting on a mantle in a cottage. When I was an undergraduate student and biology was still a joy, going to help band bats meant exploring caves and a chance to meet face to face with the objects of my fascination. Later, as a graduate student, bats offered many challenging opportunities for study and the chance to work regularly in the field – not just in the laboratory.

Today, as a teacher, I continue to learn about bats, and their study seems to make it easier for me to teach others. My students are not subjected to a steady diet of bat lore, though, for learning about bats is not an end in itself.

Bats offer a showcase of how animals interact with their surroundings and other animals, including man – a combination that makes them excellent subjects for studies designed to increase our understanding of our world. But, when all is said and rationalized, studying bats is a continuing mental and physical challenge; more than that, it can be thrilling.

Until 1977, Long-crested Free-tailed Bats had never been seen in Zimbabwe, and my excitement at finding six in a net I had stretched across a river carried me through the numbingly cold water and beyond the distractions of elephants, lions, and terrorists. Examining the contents of a net or bat trap can be better than the thrill of unopened Christmas presents.

Every scientist who works in the field accumulates a repertoire of stories from his experiences there. These stories often suggest that field work provides a setting for adventure – not always happy – and are usually retold time and again, frequently to the discomfort of other participants. Walking up a sandy riverbed in the dark, beating a hasty but dignified retreat from a herd of elephants, and finding your mouth too dry to whistle is not funny until later. But watching a colleague plunge headlong into a chest-deep mudhole is entertaining at once and the story only improves with age.

Taking a group of 10 students into the field for several weeks to study bats and caves – watching the fastidious emerge mud-coated from a rat-hole of a cave in the Canadian Shield – is an exercise in teaching and learning. At the end of the excursion, it is difficult to separate the work from the fun, and by then, everyone has a supply of anecdotes that will last for years.

So I study bats and find stimulation, the intellectual challenge of trying to decipher what they are doing, the entertainment of visiting new places or returning to old haunts, and the chance to meet interesting people. I hope that this book will convey my appreciation of bats and the importance of their fascinating world: from their world we can increase our understanding of ourselves and our planet.

Acknowledgments

I am particularly indebted to four scholars who introduced me to, and whetted my appetite for bats. The late Dr Roland E. Beschel took me into my first bat cave, and Dr R.L. (Pete) Peterson steered me into and along the path of bat research. Dr Donald R. Griffin and the late Dr Kenneth D. Roeder helped me to appreciate the diversity of bat research, and to expand my horizons.

It has been my very good fortune to have worked with a stimulating group of students and colleagues, some far, some near, who have kept the wheels turning and the ideas perking. They include, but are not limited to: Robert M.R. Barclay, Gary P. Bell, Jackie J. Belwood, Neil G.H. Boyle, David K. Cairns, George R. Carmody, Richard A. Coutts, David H.M. Cumming, James H. Fullard, Connie L. Gaudet, Judith F. Geggie, T. Michael Harrison, Robert M. Herd, G. Roy Horst, Thomas H. Kunz, Marty L. Leonard, Lynda S. Maltby, David J. Oxley, Paul A. Racey, James A. Simmons, Donald W. Thomas, Christine E. Thomson, Burleigh Trevor-Deutsch, Tracey K. Werner, Dedee P. Woodside, and Gregory C. Woodsworth. Honours research students, and others involved in field courses, animal behaviour, chordate biology, form and diversity, and introductory biology at Carleton University have also contributed to my development and I thank them all.

Jack (J.G.) Woods and the late J. Allan Graham accompanied me during long hours in the field, from oppressively hot and aromatic bat colonies to cold and damp caves and mines, and I am very grateful to them. Dr H. Gray Merriam deserves special mention as a companion in the field and a stimulating colleague.

I thank Paul Geraghty and Connie Gaudet for their artistic endeavours on my behalf, and Merlin Tuttle, Ulla Norberg, Carl Brandon, Donna Howell, Paul Racey, Gary Bell, and Donald Thomas for allowing me to use some of their excellent photographs. I also thank Robert Herd for his work in the dark-room.

Special thanks go to Doug Fischer who patiently edited the manuscript and to Judy Williams and Ian Montagnes at the University of Toronto Press who made the whole operation possible. The following people read all or parts of the manuscript and made many helpful comments and suggestions: Alvar Gustafson, James A. Simmons, Paul A. Racey, Iain F. Downie, Geoffrey L. Holroyd, Meg Cumming, and Thomas Kunz.

My research has been generously supported by the Natural Sciences and Engineering Research Council of Canada and the Faculty of Graduate Studies and Research at Carleton University. I am grateful to the Ontario Arts Council for awarding me a grant to facilitate the preparation of this manuscript.

My greatest debt is to my wife Eleanor who read different versions of the manuscript, tolerated my long absences in the field, and provided all sorts of moral support.

I must conclude by thanking the bats, all of them.

Just Bats

Introduction

In modern Western society, just the mention of bats is enough to conjure images of bloodthirsty vampires, draughty dark castles, and evil spirits. Bring up the subject at almost any gathering and the reaction is generally the same – bats are dirty and dangerous, common carriers of rabies, ugly creatures that get tangled in your hair.

Old myths die hard. For no better reason than that they are nocturnal and shy, bats are viewed with foreboding by a large segment of society; arguably, they are the most misunderstood, feared, and persecuted of mammals. But it is not that way everywhere. In some societies, bats are considered beneficial: they are symbols of long life, good luck, and fertility. The art of some of these societies shows a considerable awareness of bats. They are frequently depicted in Chinese art as decorations on clothing and utensils – sometimes stylized, other times true enough in detail to permit identification of species. Stories of bats fill the lore of many tribes from Central and South America. Interestingly, bats rarely appear in African rock paintings, and their absence from these illustrations is puzzling since the paintings are often located in rock shelters and caves used as day roosts by bats.

But like them or not, man has long been fascinated by bats, a situation that has resulted in some practical, if not positive, applications. In tropical and subtropical countries with large populations of gregarious bats, accumulations of bat guano from roosts have been used as a source of saltpetre to make gunpowder or fertilizer. Guano mined from bat caves in the southern United States was used to produce gunpowder during the War of 1812 and again during the Civil War. As recently as the Second World War, the American military entertained the possibility of using bats to carry incendiary bombs to enemy sites. The plan called for the bombs to be affixed to bats which would be released at a set altitude from cages dropped over Japan by parachute. It was hoped that the bats would select strategic roosts and chew through the cord connecting them to their destructive load, and that they would be out of the area when the bombs ignited. Military brass elected to abandon Project X-Ray after one of their own buildings was gutted by the fire generated by an errant bat-carried bomb.

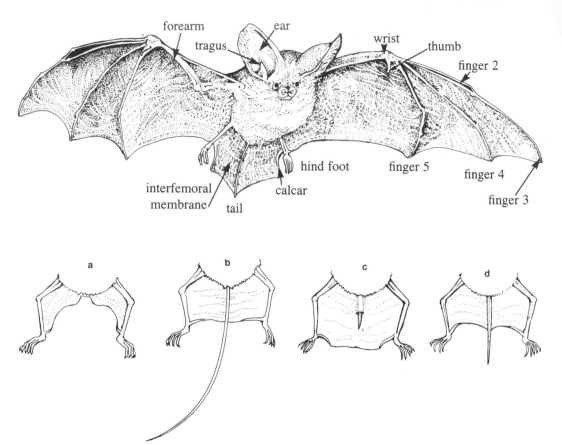

This flying Long-eared Bat provides a basic map to different parts, which are labelled. The arrangement of the tail and interfemoral membrane (where present) is useful in distinguishing between families of bats outlined in the table on pages 8 and 9 (the interfemoral membranes are here viewed from above). Some bats lack an external tail and they have interfemoral membranes modified accordingly (a). This condition is common among New World Leaf-nosed Bats and flying foxes and their relatives. Mouse-tailed Bats have long thin tails extending beyond the ends of their intefemoral membranes (b), while Sheath-tailed Bats have tails protruding through the membrane (c). In Free-tailed Bats the tail is thick and extends beyond the end of the membrane (d). In many other families, such as Plain-nosed Bats, the tail is fully enclosed by the interfemoral membrane as shown here for the Long-eared Bat, although one or two vertebrae may be free of the membrane. Drawing by Connie L. Gaudet.

Why the fascination with bats? Certainly flight and nocturnal habits are major contributors, and our relatively recent discovery of echolocation (bats' ability to listen to the echoes of their own signals to detect obstacles) has only added to the mystique. Bats are also common, occurring almost everywhere in the world except for most of the Arctic and all of the Antarctic. Some species live on remote oceanic islands (the Hoary Bats of Hawaii, for example), and there are records of Red Bats on Southampton Island in the Canadian Arctic. Other species of bats occur in Tierra del Fuego at the other end of the Americas. There are more than 850 species of bats, placing them second only to the Rodentia (mice, squirrels, marmots, etc.) as the most diverse of the mammals. In many tropical parts of the world, there are more bats than any group of mammals, and even in Canada, a northern temperate country, 19 species have been reported. Although bats are most diverse in South and Central America, nowhere is their dominance of the mammal fauna more striking than in Africa. Although Africa usually conjures up visions of larger mammals – lions, elephants, giraffes, and antelopes – almost every part of the continent features more species of bats than any other group of mammals, including rodents.

The basic design of all bats is uniform enough to allow anyone to recognize them as such. The name for the order, Chiroptera, reflects the structure of the wings – folds of skin stretched from the sides of the body to elongated finger bones (chiro – hand; ptera – wing). In some bats, the Bare-backed Fruit-eaters from the East Indies and the Naked-backed Moustache types from South and Central America, the wing membranes meet on the middle of the back. In all species, the thumbs are relatively free of the wing membrane, which is divided into separate compartments by the elongated fingers. In this way, bats differ from the extinct flying reptiles, the pterosaurs, whose wings were also folds of skin, but supported by a single elongated finger. Bats' hind legs also support the wing membranes, and in species with a tail it is wholly or partly enclosed by the membrane extending between the hind legs. Some bats lack an external tail and do not have well-developed membranes between their hind legs.

All bats have eyes and can see to varying degrees. There is also considerable variation in the design of the face and ears. Some species have simple ears with no special features, others have enormous ears with various embellishments; similarly, some have relatively simple faces and others grotesque facial ornaments with leaf-like structures. It is thought that ornamentation of the ears and face is related to echolocation. The nostrils of some bats are tubular, but their function is not clear. East Indian fruit bats with tubular nostrils might use them as snorkels, permitting them to breathe while the face is buried in a ripe, juicy fruit. But some insectivorous bats from the same area also have these nostrils and the snorkel explanation does not seem to apply.

The dog-like face, simple ear, and claw on its second finger are characteristics identifying this Gambian Epauletted Fruit Bat (100 grams) as a relative of the flying foxes (Pteropodidae); see table on pages 8 and 9.

Biologists are still not sure of the functional significance of the tubular nostrils of this Lesser Tube-nosed Bat (30 grams) from Papua New Guinea. The claw on the second finger identifies it as a relative of the flying foxes.

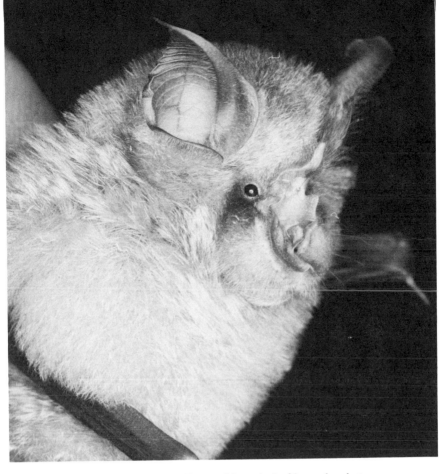

The nose-leaf of the Bushveld Bat (7 grams) is typical of horseshoe bats (Rhinolophidae), which occur widely in the Old World, from Europe and Asia to Africa and Australia. This African species often roosts in caves.

Bats are divided into two main groups, the Megachiroptera and Microchiroptera. Megachiroptera are generally the larger bats, as the prefix implies, and include about 150 species characterized by large eyes, simple ears, and, in most cases, no echolocation ability. There are about 700 species of Microchiroptera, most with small eyes, complex ears, and the capacity to echolocate. Although most biologists believe the two groups are closely related, some have suggested that they evolved from different ancestors. In Australia, for instance, it is common for people to distinguish between flying foxes (Megachiroptera) and bats. Within the two suborders, species are grouped into 17 or 18 families, and members of the same family share common characteristics such as free-tails, sheath-tails, or plain-noses. A detailed list of bat families is presented in the accompanying table.

TABLE   The living families of bats, their distribution, diet, and distinctive features

Family name	Distribution	Number of species
Flying Foxes, Old World Fruit Bats (Pteropodidae)[1]	Old World[2] tropics, Africa, Asia, Australia	150
Mouse-tailed Bats (Rhinopomatidae)	North Africa east through southern Asia to Sumatra	2
Sheath-tailed Bats (Emballonuridae)	Old World and New World[3] tropics	44
Butterfly Bats (Craseonycteridae)	Southeast Asia	1
Bulldog Bats (Noctilionidae)	New World tropics	2
Slit-faced Bats (Nycteridae)	Africa, except the Saharan region, Madagascar, Malayan Peninsula, Sumatra, Java, Borneo and adjacent islands	13
False Vampire Bats (Megadermatidae)	Africa, India, southeast Asia, East Indies and Australia	5
Horseshoe Bats (Rhinolophidae)	The Old World, from temperate to tropical areas	69
Old World Leaf-nosed Bats (Hipposideridae)	Old World Tropics	59
Moustache Bats (Mormoopidae)	New World Tropics	8
New World Leaf-nosed Bats (Phyllostomatidae)[4]	New World Tropics	126
Funnel-eared Bats (Natalidae)	New World Tropics	4
Thumbless Bats (Furipteridae)	New World Tropics	2
New World Disc-winged Bats (Thyropteridae)	New World Tropics	2
Old World Disc-winged Bats (Myzopodidae)	Madagascar	1
Plain-nosed Bats (Vespertilionidae)	Almost everywhere in the world where you find bats	283
New Zealand Short-tailed Bats (Mystacinidae)	New Zealand	1
Free-tailed Bats (Molossidae)	New and Old World tropics and subtropics, with some species occurring in temperate areas	82

[1]Denotes the Megachiroptera; all other families are in the Microchiroptera.
[2]Old World = Eurasia, Africa, Australia
[3]New World = North, Central, and South America
[4]Includes three species of vampire bats, sometimes treated as a separate family, the Desmodontidae.

Diet	Distinguishing features
fruit, nectar, and pollen	dog-like face; claw on second finger in most species; large eyes, small ears (p. 6, top)
insects	long, slender mouse-like tail protruding from the end of the interfemoral membrane (p. 4, b)
insects	tail protruding through the interfemoral membrane (p. 4, c)
insects	tiny bats with no external tail
insects and fish	enlarged hind feet for catching fish (p. 54)
insects, other arthropods, and vertebrates	large ears; tail ending in a 'T-shaped cartilage; deep slit in the middle of the face (p. 47)
animals from insects to mice, fish, bats, and birds	no upper incisors; prominent nose-leaf (p. 37)
insects	distinctive horseshoe-like nose-leaf (p. 7)
insects	distinctive nose-leaf, often decorated with club-like projections (p. 11, top)
insects	distinctive moustached face, no prominent nose-leaf (p. 49)
insects, fruit, nectar and pollen, vertebrates, and blood	most species with distinctive spear-like nose-leaf (p. 50)
insects	funnel-shaped ears
insects	thumb very reduced in size, almost vestigial
insects	adhesive discs on wrists and ankles (p. 89, a and b)
insects	adhesive discs on wrists and ankles; fundamental design differs from New World counterparts (p. 89, c)
insects, other arthropods, and fish	no nose-leaf; tail fully enclosed by interfemoral membrane (pp. 10, top. 34, 41, 90, 137)
insects, fruit, nectar, and pollen	needle-sharp claws and a short tail
insects	thick tail extending beyond the end of the interfemoral membrane (p. 4, d)

This Long-legged Myotis (8 grams) from British Columbia is a typical plain-nosed bat. Note the absence of a nose-leaf. The well-developed tragus (protruding from its ear) assists the bat in vertical localization of targets during echolocation.

Ansorg's Free-tailed Bat (12 grams) has the wrinkled lips typical of many Free-tailed Bats. This is an insectivorous species which often flies at considerable heights above the ground while hunting. This species is widespread in Africa.

Great Round-nose-leafed Bats (30 grams) occur in Southeast Asia. Their nose-leaf is typical of most Old World Leaf-nosed Bats (Hipposideridae) and serves to distinguish them from other bats with facial ornamentation, notably Horseshoe Bats (Rhinolophidae). Sketch by Paul Geraghty.

This Mauritian Tomb Bat (25 grams) is representative of the Sheath-tailed Bats (Emballonuridae). Mauritian Tomb Bats are widespread in Africa where they usually roost in the open on the trunks of trees. Other Sheath-tailed Bats are found throughout the tropics. This species produces an echolocation call which is audible to some human observers. Sketch by Paul Geraghty.

Unlike birds, bats do not have uniformly accepted common names, probably because their nocturnal habits make them unfamiliar to the public. In preparing this book, I have tried to use the appropriate common names and have presented the scientific equivalents in an appendix on page 153. This approach is used in sympathy with public objections to scientific names, which are often unwieldy and unfamiliar. For some bats, however, the argument for common names is moot; neither the 'common' Duke of Abruzzi's Free-tailed Bat nor the scientific *Tadarida aloysiisabaudiae* rolls easily off the tongue.

Because they fly, bats are often likened to birds, and in comparison they suffer. Birds have exploited a number of lifestyles not developed by bats, most notably swimming, diving, and running. There are also more than 8,000 species of birds, almost 10 times the number of bat species. One reason for the greater diversity of birds could be their age. Fossils of the first bat, *Icaronycteris index*, are about 60 million years old, while the oldest bird fossils, *Archaeopteryx lithographica*, date back 135 million years. Although there is some dispute among paleontologists about this early bird's ability to fly, the fossils clearly show that the evolutionary development of birds had a significant head start on that of bats. Perhaps a more important reason for the greater diversity of birds is anatomical. The hind legs of birds are not part of the basic wing structure; they thus have a terrestrial flexibility not available to bats, whose hind legs are an important component of wing support.

Some scientists believe the ancestor of bats was a small shrew-like animal that ran about in trees chasing insect prey. It has been suggested that this pre-bat had elongated fingers supporting webs of skin extending from its body, and that these webbed appendages were used as parachutes and nets to catch insects. I suspect that this hypothetical ancestor located insects by echolocation, which permitted the animal to catch flying insects at night. If this hypothesis is correct, echolocation and the ability to catch prey in darkness were probably the key to the success of bats; their ancestors must have had some advantage over birds of the time.

Unravelling the mystery of the origin and evolution of bats will only be possible with more fossils to study. But finding them may be more a function of luck than planning, since bats, small animals with delicate bones, are not well represented in the fossil record. Recognition of bat fossils is also not easy. It is possible that some teeth now sitting in museum collections and labelled 'shrew' actually belonged to an ancestral bat.

But whatever the evolutionary recipe, it was successful – the modern descendants are widespread and numerous, and have broadened their food habits to include fruit, nectar and pollen, scorpions, fish, birds, mice, frogs, blood, and even other bats.

Most bats are small. The flying foxes, which weigh up to a kilogram and have wing-spans of nearly two metres, are the largest. But these are the Goliaths of the bat world and only a few species achieve such dimensions. At the other extreme are the Butterfly Bats of Southeast Asia, which weigh two or three grams and have wing-spans of about 20 centimetres. The world's largest insectivorous bat, the Giant Leaf-nosed Bat from Africa, weighs 150 grams, but fewer than 50 species of Microchiroptera weigh more than 50 grams as adults.

In recent years, hundreds of scientific papers dealing with bats have been published annually. These are in addition to the many books which report basic information about bats. Faced with this kaleidoscope of information, one might think there is little left to learn about bats. Nothing could be further from the truth; in fact, one of the main themes of this book is how much remains to be understood about them. To provide a sampling of readings about bats, I have included a list of selected references as a basis for further exploration.

Several important steps in the study of bats have led to this explosion of scientific information. New discoveries about echolocation play a significant role, inspiring many further studies on topics including brain anatomy and information-processing in the brain. However, several important developments in methods of studying bats have also been key factors in this information boom.

One limitation to the study of bats is their capture. Japanese mist or fowling nets made of fine nylon mesh have been used to catch birds and bats, previously captured only in their roosts or by prudent shooting. But despite many years of sampling with these nets, we know several species of bats only from single specimens, and, in many cases, these were strays caught by accident. Mist nets also have the disadvantage of working by entangling their victims, so that each one must be untangled, often a time-consuming operation.

This untangling presents some practical problems to biologists trying to sample bats leaving a cave to feed at night, particularly if thousands of bats fly out of the cave at once. In such cases, harp traps, made from wires strung vertically on a harp-like frame, have been used successfully. The departing bats fly into the strings and slide down into a waiting container, saving the time and trouble of untangling them from nets. In 1974, Merlin Tuttle described a free-standing harp trap he had developed while a graduate student at the University of Kansas. The 'Tuttle Trap' has a pair of harp-like frames to catch the bats, and beneath the frames is a bag into which the captives fall and from which it is easy to remove and process them quickly. By minimizing the time bats are held before release, and presumably their trauma, the Tuttle Trap has become a cornerstone in the development of bat study.

It might seem logical that the capture of bats that can echolocate to find insects as small as midges (a diameter of less than one millimetre; see page 26) would be impossible. With many species, however, this is not the case. Donald R. Griffin, an American biologist and a pioneer in the study of echolocation, labelled this phenomenon 'the Andrea Dorea effect,' alluding to the disastrous 1956 collision between the Italian luxury liner and another vessel, both of which were equipped with radar designed to avoid such a tragedy. Similarly, bats do not always heed the information in the echoes of their calls, and in this way are like people who bump into walls or doors that they 'saw' but did not 'notice.'

Bats are very sensitive to disturbances around their roosts, a fact which imposes further limitations on those wanting to study them. Animals captured or disturbed inside their roosts often abandon these sites, making continued study impossible or at least very difficult. A night of netting or trapping in an area is usually enough to condition local bats to the perils of the site. Television cameras and starlight scopes functioning under low light conditions permit

A Tuttle Trap, a double-framed harp-like structure, is an important innovation in the capture of bats. This trap is set in the gorge of the Sengwa River in Zimbabwe, where it caught several Lesser Yellow House Bats and two Hildebrandt's Horseshoe Bats. The trap frame is 1.8 by 1.5 metres.

observation in near darkness and allow biologists to spy on bats in their roosts and as they forage.

Techniques and technology for observation of bats in their roosts have proceeded hand in hand with biologists' ability to study them away from their roosts. For instance, the application of small celluloid bands coated with coloured reflective tape to the wrists of bats allows subsequent identification by flashing a light on the band and looking for the coloured reflection. However, reflective bands mean the bat must be captured in the beam of light, a disruption that often changes its behaviour. A more progressive technique has been developed by another American biologist, Edward Buchler, who used the fluid from Cyalume[K] light sticks to make small lights he attached to the bats. 'Buchler Tags' make it easy to follow the flight of a lighted subject, and marked bats usually go about their business relatively undisturbed. I find bats that have not fed prior to light-tagging are more co-operative than those already sated: the former spend relatively little time trying to remove their tags before going on to feed, but the latter retire to a roost where they industriously try to groom off the light. Buchler Tags last less than 12 hours, however, and are not suitable for the biologist trying to follow the same bat for a longer period.

A 0.9-gram transmitter has been glued to the fur on the back of a Spotted Bat. Signals from the transmitter allowed us to follow the animal as it moved about on its nightly foraging. These signals were detectable at about one kilometre.

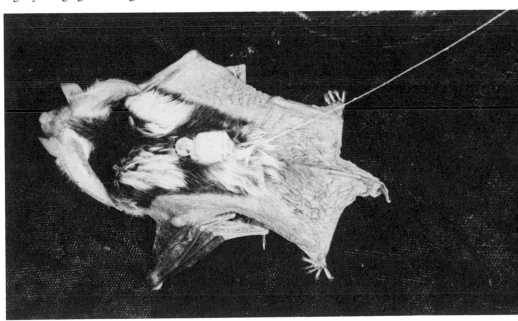

For long-term studies, many biologists rely on small radio transmitters attached to their study animals. Developments in electronics have reduced the size of the radio tags, some with a range of over one kilometre and a battery life of three weeks, to less than one gram. Most available transmitters are larger, though, and because biologists try to use tags weighing less than 10 per cent of the animal, the majority of bats are still too small for even the smallest radio tags. Nevertheless, radio tags have already produced considerable information on the behaviour of bats and promise to make more contributions as their weight and bulk are further reduced.

Developments in electronics have also promoted the study of other aspects of bat biology, particularly echolocation. When Griffin first went into the field in the early 1950s to record the ultrasonic echolocation calls of bats, the apparatus he needed filled the back of a pick-up truck; modern versions of the same equipment easily fit into the trunk of a sub-compact car with room to spare. One combination of microphone, analyser, and portable oscilloscope permitting an in-field view of bat echolocation calls weighs less than three kilograms and can conveniently be carried around the neck.

Because of these technological improvements, and the increasing number of biologists striving to learn more about bats, we are on the threshold of many exciting advances in the study of these obscure animals. Some of this research seems to hold little benefit for man beyond providing a better understanding of our world. However, studies of echolocation promise to help us unravel the secrets of how brains process information, and they offer some assistance for the development of an acoustic orientation system for the blind. Similarly, studies of the feeding behaviour of bats may also lead to their use in operations designed to control populations of harmful insects.

Certainly, such beneficial and fascinating animals deserve a better fate than fear and persecution.

Flight

The ability of bats to fly makes them different from all other mammals. Bats have gone beyond the gliding and parachuting abilities of several lineages of flying squirrels, flying lemurs, and Australian gliders, and developed active flight.

Unravelling the details of bat flight was initially achieved by comparative anatomists who studied bone structure and the arrangement and relative sizes of muscles to predict how the system worked. More recently, innovations in high-speed photography have permitted further dissection of flight movements, and when these stop-action pictures are used in combination with advances in electromyography – the insertion of small electrodes into specific muscles   it is possible to assemble an integrated portrait of how bats fly.

Because bats have developed active flight, their wings provide lift and propulsion, the essential components of flight, along with flight speed, the angle at which the wing moves through the air (the angle of attack), the design of the wing, and the tautness of the wing membrane. The amount of lift, for example, depends partly on the flight speed and the angle of attack. At slower speeds or steeper angles of attack, the amount of lift is drastically reduced and can result in stalling. Moreover, different designs of wings have different stalling speeds, whether they be the wings of bats, birds, or aircraft.

Lift is provided by the airfoil section of the wing membranes between the side of the body and the fifth finger; propulsion is generated by the distal part of the wing, from the fifth finger to the tip. During flight, there is little change in the angle of attack of the lift-providing membranes, but drastic alterations in the angle of the power-producing parts of the wing. This is because the wing-tips move faster and through a greater arc than parts of the wing closer to the body.

The power for propulsion is produced by the downstroke of the wing; it follows that, because they do more work, the muscles producing the down-stroke are much larger than those responsible for the upstroke. Anyone who has carved a fowl or eaten fried chicken is familiar with the flight muscles of birds – the breast meat. Like birds, bats power their downstroke by contractions of muscles in the chest; but the number and arrangement of flight muscles, particularly those involved in the upstroke, are fundamentally different.

Bird flight is powered by two pairs of breast muscles – one for the down-stroke, the other for the upstroke. Both sets are on the breast, although only the muscles that generate the downstroke do so directly. The upstroke requires much less power and is achieved by contraction of muscles operating by a simple pulley system made up of a tendon which runs through the shoulder girdle and attaches to the humerus or upper arm bone. Birds also have promi-nent keels on their breastbones providing a large surface area to which the flight muscles attach.

In bats the downstroke is powered by three main pairs of muscles located on the breast. Two pairs pull the wing down in the way people use similar pairs of muscles to bring arms outstretched on either side together in front of them. The third pair is involved in the downstroke through an arrangement in which the shoulder blade (scapula) locks with the top end of the humerus and rocks about its long axis. This arrangement adds a functional joint to the wing and works on the same principle as a throwing stick.

Three similar pairs of muscles located on the bat's back elevate the wings, returning them to the upward position in preparation for another downstroke. Several species, notably some free-tailed and plain-nosed bats, have a 'door-stop' arrangement between the humerus and scapula that automatically stops the wing at the top of the upstroke, thereby avoiding the need for muscular action to assist with this.

Unlike birds, bats do not have a well-developed keel on the breastbone, and this, along with the top and bottom arrangement of flight muscles, gives them a relatively shallow profile through the chest. Bats can thus roost in narrow crevices or squeeze through small openings into more roomy roosts.

Because bats have teeth, they face a problem of weight concentration in the front end, much as a paper airplane is weighed down by a paper clip in its nose. Bats compensate for this problem by having short necks, which help to keep the weight concentrated towards the centre of the body. Birds grind their food with gizzards, which are located close to the centre of the body, thereby avoiding the problems of front-end loading. In both birds and bats, food usually passes quickly through the digestive tract, reducing the amount of time the weight of undigested food is carried about. Little Brown Bats, for instance, generally pass food within 20 minutes, about the same length of time as birds of similar size.

Birds avoid another weight problem by laying eggs. Giving birth to live young means that female bats must carry an extra burden during the late stages of pregnancy, a period which in some species may be as long as one month. However, live birth means that bats do not have to build nests, especially since baby bats can hang up on their own from birth.

Ulla Norberg took this spectacular set of photographs of a flying Long-eared Bat (10 grams) during her study of its flight mechanics. The three photographs provide three different simultaneous views of the flying bat.

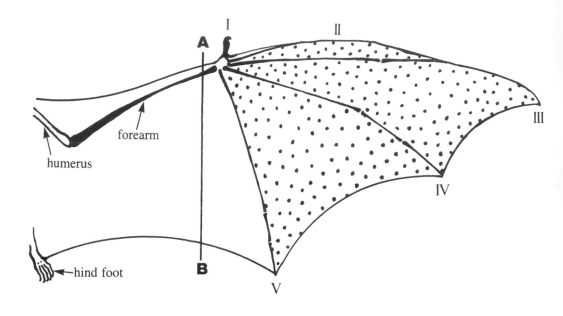

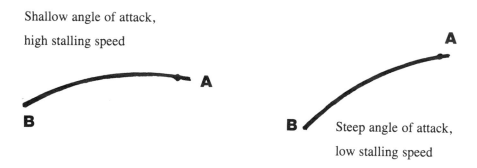

Shallow angle of attack,
high stalling speed

Steep angle of attack,
low stalling speed

The wing of a Red Bat (actual size) shows the upper arm (humerus), forearm, thumb (digit I), the four fingers (II, III, IV, and V), and the hind foot. The forearm is about 39 millimetres long. Air moving more rapidly over the upper curved surface than under the wing creates a negative pressure above the wing surface, producing lift. During flight, the parts of the membrane that produce propulsion are indicated by the stippled area; those producing lift by the clear area. The cross section (A-B) shows the airfoil section of the wing at shallow, and steep, angles of attack.

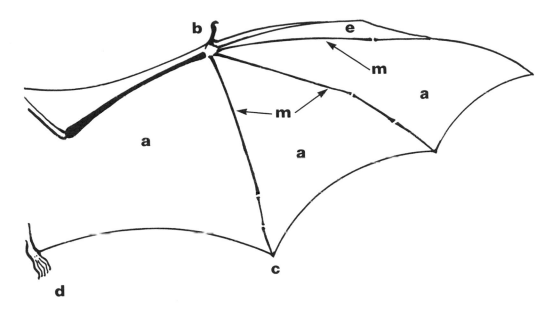

The same Red Bat's wing shows how different components contribute to the
development of lift, propulsion, and a rigid leading edge to the wing during flight.
The membrane (a), wrist (b), fifth digit (c), and hind limb (d) maintain the airfoil
section and produce lift during flight. Within the membranes are elastic fibres to keep
it taut; some species, notably Free-tailed Bats, also have sheets of muscle in the
membrane. In the wrist, the metacarpal bones (m) dovetail with an extra bone (called
a pisiform bone) to provide sufficient bracing. A muscle, the abductor pollicus
longus, pulls one of the wrist bones (the scaphoid) towards the forearm. The fifth
digit is 'I'- or 'T'-shaped in cross section to help the wing resist upward air pressure.
The hind limb can be moved to increase tension on the membrane and to change the
airfoil section of the wing. In some species, components of the wing providing
propulsion account for less than 20 per cent of the wing area. Modifications in the
wrist and a muscle (the extensor carpi radialis longus) assist in maintaining rigidity of
the membrane between digits II and III (e), reducing resistance and contributing to an
effective leading edge of the wing.

Perhaps birds' biggest flight advantage over bats is feathers. Feathers provide insulation by helping to keep heat in or deflect it away, and bats rely on fur to accomplish these same ends. Some biologists have suggested that feathers evolved partly to shield birds from sunlight, while bats, whose ancestors were probably nocturnal, required fur to keep heat in. Feathers are probably largely responsible for making birds more effective than bats in many facets of flight. The slotting gaps between the wing feathers of birds provide the capacity for extra lift. Furthermore, feathers are extremely light-weight and resistant to damage. Yet although bat wings are delicate, they are also relatively strong and resilient. The division of the wing membranes into different compartments bordered by finger bones makes bats less vulnerable to immobilization if a membrane is torn or a supporting finger bone is broken. Tears in the wing membrane, which is basically a fold of skin enclosing blood vessels, some connective tissue, and varying amounts of muscle, also tend to heal quickly. Studies of Pallid Bats, for instance, showed that holes 14 millimetres in diameter healed completely in just over three weeks.

But despite the additional weight created by teeth and, in the case of pregnant females, developing embryos, some bats are capable of very manoeuvrable flight. One of the most impressive demonstrations I have witnessed was provided by a Dent's Horseshoe Bat that was released into a closet one metre wide, 80 centimetres deep, and 30 centimetres high. The bat, which had a wing-span of about 25 centimetres, flew at length without touching the walls, ceiling, or floor.

By altering the angle of their wings, bats are able to roll or slip to one side, two manoeuvres that involve loss of altitude and are usually associated with tight, downward turns and attempts to catch insects. Turning is achieved by sideslipping or by beating one wing faster than the other. Some bats regularly glide in flight, but none is as accomplished at it as some birds.

In flight, the tails of bats and the membranes that enclose them act to alleviate swayings of the body, and may also contribute substantially to total lift and manoeuvrability. But since many species do not have tails and interfemoral membranes, it follows that they are not essential to flight.

As might be expected, insect-eating species are the most skilful fliers among bats. The flight of flying foxes, consumers of fruit, pollen, and nectar, is usually laboured and clumsy, reflecting less developed anatomical specializations for flight. Conversely, the connection of shoulder blade and the top end of the humerus in free-tailed bats indicates they are among the species most specialized for rapid flight.

Wing shape and the weight of the bat in relation to its wing area also affect flight ability and influence hunting behaviour. Small bats with very broad wings tend to be more manoeuvrable in flight than larger, broad-winged species because less lift is required to move their lighter bodies. And because they can

A stroboscopic view of a flying Egyptian Slit-faced Bat. Photograph taken by
Gary Bell.

develop greater lift, light, broad-winged bats also have lower stalling speeds
than species with longer, narrower wings. However, while species with narrower
wings have higher stalling speeds, they are also capable of faster flight. Free-
tailed species such as the Greater Mastiff Bat must roost where they can initiate
their evening flight with a free vertical drop of several metres. This initial
free-fall permits them to build their air speed above their stalling speed and to
begin horizontal flight.

Bats' wings serve a number of functions besides transportation. Many insectivorous species use their wing membranes to catch their prey, an advantage
that allows them to counter last-minute evasive manoeuvres by the insects. On
hot days, flying foxes resting in exposed locations use their wings as fans, and
wings are used by males in some mating displays to attract females. The wings
also act as a sort of radiator to disperse heat built up by flight.

Studies of bats flying in wind tunnels show their rates of energy consumption
are comparable to similar-sized birds. While it is true bats consume fuel up to
three times as fast as the highest rates recorded from small terrestrial mammals,
a 90-gram bat uses less than one quarter of the energy consumed by a terrestrial
animal travelling an equal distance.

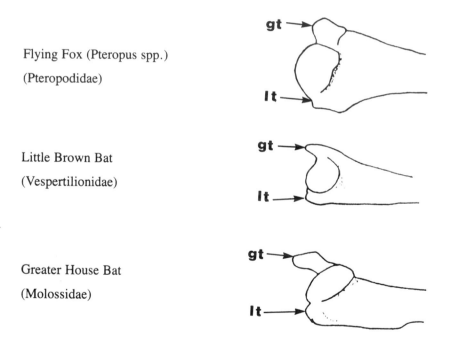

Flying Fox (Pteropus spp.)

(Pteropodidae)

Little Brown Bat

(Vespertilionidae)

Greater House Bat

(Molossidae)

The top ends of the humeri (upper arm bones) of three bats are compared in this
diagram to demonstrate the increasing development of the greater tuberosity (gt), a
projection which acts as a doorstop to halt the upstroke of the wings. This doorstop
works against the scapula or shoulder blade. The other side of the humerus bears the
lesser tuberosity (lt), which anchors one of the muscles responsible for extending the
forearm.

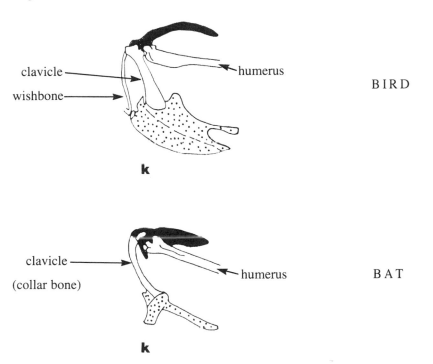

The shoulder girdles of a bat and a bird are compared in side view, with the breastbones (sterna) indicated by the stippled area and the shoulder blades (scapulae) in black. Flying birds have an extensive keel (k) on the sternum that is not well developed in bats. In birds, two pairs of muscles powering the downstroke and upstroke are both located on the breast. In bats, however, the downstroke is powered by muscles on the breast, while the upstroke muscles are on the back.

In addition to advancing studies of the energy consumption of bats in flight, wind tunnels have also been important tools in researching bat aerodynamics. Air is passed through the tunnel by fans, which can be adjusted to manipulate air speed. A flight chamber built into the tunnel is screened at either end to confine the subject, and the walls and ceiling of the chamber are usually made of glass or clear plastic to permit an unobstructed view. The bat remains stationary, changing the number of wing beats to counter changes in air speed, allowing a detailed examination of the mechanics of flight. The rate at which the bat uses oxygen at different speeds is measured by placing a mask over its face.

Flight makes bats conspicuous, gives them mobility to exploit dispersed food and roost resources, and accounts for their wide distribution in the world.

Echolocation

Man has long marvelled at the ability of bats to manoeuvre in total darkness, for under these circumstances a Barn Owl, also a nocturnal creature, blunders into objects in its path.

In the late 1700s, Lazarro Spallanzani, an Italian scientist, conducted experiments to determine how bats accomplish orientation in darkness. His studies, which included efforts to deny his experimental bats their use of smell, touch, and vision, showed that a bat lost its powers of orientation when its head was placed in a sack. Spallanzani concluded that his bats had a 'sixth sense,' and his results prompted Charles Jurine, a Swiss surgeon, to perform a series of his own experiments. Jurine reported in 1794 that if a bat's ears were blocked it lost its ability to manoeuvre in darkness. His work renewed Spallanzani's interest, and taking Jurine's work a step further, the Italian found that when he inserted brass tubes into a bat's ears, it could orient only when the tubes were open; if he blocked the tubes, the bat blundered about helplessly. He deduced that ear canals were an essential part of orientation for bats, leading to the suggestion they somehow see with their ears.

Other workers who duplicated Spallanzani's experiments agreed with his conclusion. In 1920, H. Hartridge, an English physiologist, took the idea a step further by suggesting that bats might listen to sounds to detect obstacles. But it was not until the 1930s, when echolocation pioneer Donald R. Griffin, then an undergraduate student at Harvard University, took some bats to the laboratory of a professor who had a microphone sensitive to ultrasonic sound, that this suggestion was made credible. (Sounds above our range of hearing, commonly considered to be 20 kilohertz, are arbitrarily classified as ultrasonic.) Griffin used an ingenious battery of experiments, including some that had been designed by Spallanzani, to show that his bats used the *echoes* of their calls to *locate* obstacles. He coined the term 'echolocation' to describe this behaviour. For a detailed treatment of some of the work that led to man's discovery of echolocation, I recommend Griffin's book *Listening in the Dark*.

We know that some mammals, including dolphins, porpoises, several types of shrews, and even some blind people, are able to echolocate. Oilbirds from

South America and cave swiftlets, whose nests are used to make birds' nest soup, employ their echolocation calls to find their way into caves where they nest. However, a Barn Owl tracking a mouse does not use echolocation, but rather listens for sounds of movement from its prey.

Although any bat encountered in Europe, Canada, and the United States usually uses echolocation to detect obstacles and insect prey, not all bats echolocate, nor do they all use the same approach to echolocation. Most of the flying foxes do not echolocate, the several species of Dog-faced Bats found from Africa east to New Guinea being the exception. But whereas other echolocating bats produce their orientation sounds in their voice boxes, Dog-faced Bats click their tongues.

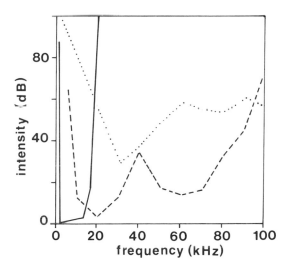

Hearing sensitivity thresholds at different frequencies are compared for man (———), a Big Brown Bat (------), and an African moth (.......). Man is more sensitive to sounds below 15 kilohertz (kHz) than a Big Brown Bat or a moth, but both bat and moth can hear ultrasonic sound (over 20 kHz) that is inaudible to humans. The curves, technically called audiograms, illustrate that at some frequencies a signal must be of greater intensity (measured in decibels, dB) to be heard. For example, at 10 kHz, humans can hear sounds of less than 5 dB intensity; however, this sound would be inaudible to the Big Brown Bat unless the intensity was increased to almost 20 dB. For the moth to hear this signal it would have to increase in intensity to 80 dB. These differences are more impressive when one remembers that decibels are a log scale of measurement.

It is important to realize that bats produce many vocalizations designed for communication (see page 134), and the squeaks and squawks of a bat in the hand or a roost are not echolocation calls. To echolocate an animal must produce appropriate sounds and be able to hear echoes of these sounds rebounding from objects in its path. Detection of large obstacles such as cliffs or walls may be easily accomplished with low-frequency sounds (long wavelengths), but smaller obstacles are only detectable using sounds of higher frequency (shorter wavelengths). Thus, the echolocation calls of Oilbirds and cave swiftlets are low in frequency and clearly audible to humans. The frequencies used by bats are higher and cover a range from about 10 kilohertz to more than 200 kilohertz.

To measure the distance to a target, it is believed bats compute the time elapsed between sound production and echo return, basing the distance on the speed of sound. The details of the target (shape, texture, movement, etc.) are probably determined by comparing the frequencies of the original call with those of the echo. The bat likely broadens the area from which it collects information by moving its head. These statements must be qualified, for we still do not fully comprehend how bats analyse sounds, and our understanding of the importance of head positions is slight. When you remember that the echo from a target one metre away arrives back at the bat's ears within six one-thousands of a second (six milliseconds) of making the sound, and bear in mind that both bat and target are often moving, it becomes obvious that echolocation is a sophisticated means of orientation.

The best way to appreciate bat echolocation is to eavesdrop with a microphone sensitive to their calls. A pond or stream on most summer evenings, a swarm of insects around a street-light, the area beside a barn used as a roost – each is a seemingly quiet locale until you can hear ultrasonic sound. One striking fact revealed by eavesdropping is the varying rate at which bats produce their sounds. As a bat approaches an obstacle, it dramatically increases the rapidity of its calls; thus a bat commuting from its roost to a feeding area may produce 5 calls each second, but the same individual trying to catch an insect increases its rate of pulse production to more than 200 per second. These bursts of pulses are called 'feeding buzzes,' although the label is not entirely appropriate because the burst occurs before the bat makes contact with its food, and biologists still have no way of determining which buzzes represent successful attacks.

The eavesdropper soon becomes aware that his bat detector is not equally sensitive to all bats, a fact recorded by Griffin when he tried to pick up the signals of some South American leaf-nosed bats. He initially thought the bats did not produce echolocation calls, but closer examination showed their calls were much less intense than those of bats he had studied previously. Griffin

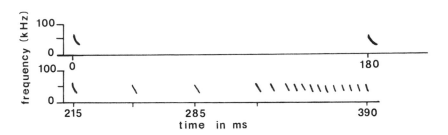

This graph shows the sequence and timing of the echolocation calls produced by a Big Brown Bat as it searched for, detected, and attacked a flying June beetle. The sequence terminates in a 'feeding buzz.' Note that the individual calls (shown as sonagrams) start at higher and sweep to lower frequencies (the significance of this is shown in the illustration on page 31). Shorter calls are produced at a faster rate as the bat closes with its prey. It is impressive to note that the time scale is in thousandths of a second (milliseconds, ms) and that the entire sequence lasted less than half a second.

labelled these species 'whispering bats.' The Little Brown Bat, the subject of Griffin's early work, produces high-intensity echolocation sounds easily picked up by a bat detector at distances of 10 to 15 metres when the bat is facing the microphone. If you held an activated smoke detector 10 centimetres from your ear, the signal would have about the same intensity as the call of a Little Brown Bat an equal distance away. Of course, because of the characteristics of your ear, you hear the smoke detector and not the bat. The call of a whispering bat, however, would have the same intensity as someone whispering in your ear from 10 centimetres.

The calls of most of the bats of the world's temperate regions are high-intensity, but a few species produce signals of intermediate strength. In more tropical regions, there are larger numbers of species producing low-intensity calls. Because the power of a bat's call influences the distance at which it can detect targets, it is not surprising to find that many high-intensity bats have a better ability to detect echoes from distant targets than species using calls of intermediate or low intensity.

If your bat detector is tunable through a range of frequencies, you will also be able to observe that echolocating bats do not rely on one frequency. Instead, they use frequency modulated (FM) calls, signals that start at one frequency and sweep down to another. For instance, Big Brown Bats use calls that sweep from 80 to 30 kilohertz, while those of Little Brown Bats go from 100 to 40 kilohertz.

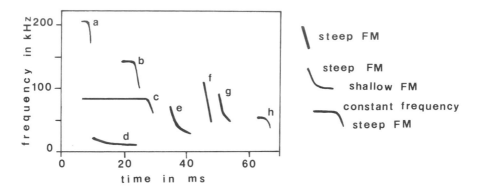

The patterns of echolocation calls produced by some bats are revealed in these sound pictures or sonagrams. Bats use calls of different durations and a wide range of frequencies which change over the duration of the call. This makes it possible to distinguish different species of bats by their calls in the same way different birds can be recognized by their songs. The basic components of echolocation calls, frequency modulated (FM) sweeps or constant frequency (CF) segments, are shown at the right. Frequency modulated calls may involve a steep or shallow sweep, defined by the frequencies covered and the timing of the sweep. Depicted here are the echolocation calls produced by the following bats as they searched for insect targets in the field: African Trident-nosed Bat (a), Noack's African Leaf-nosed Bat (b), Greater Horseshoe Bat (c), Martienssen's Free-tailed Bat (d), Big Brown Bat (e), Mexican Long-eared Bat (f), Little Brown Bat (g), and Mexican Bulldog Bat (h).

(Opposite) This illustration demonstrates the effect of echolocation call design on the ability of bats to locate targets. Increasingly accurate information about the target's location (stars) and details of its structure are provided as the signals change from constant frequency (CF) to steep frequency modulated (FM). The addition of a harmonic or overtone broadens the bandwidth of the signal (the number of frequencies involved) to further increase acuity. The position of the target as perceived by the bat is shown as the stippled area.

Some radio signals are also frequency modulated, but these are continuous, whereas the FM calls of bats are discrete pulses of sound. Some bats, notably several horseshoe and Old World leaf-nosed bats and one species of moustache bat, use a constant frequency or pure tone component, followed by a brief FM sweep. Besides varying the frequencies of any one call, bats can also add or delete overtones or harmonics.

Laboratory studies in which bats have been challenged to make decisions about targets have allowed biologists to appreciate how calls are used to gather information. A Big Brown Bat searching for a June beetle uses one kind of call, but quickly changes the design of the signal once it has found a target and starts to close in on it. The FM part of the call probably tells the bat about the texture of the target, and its position in horizontal and vertical space; harmonics further assist in pinpointing its location.

A significant potential source of error for echolocating bats is the Doppler effect. Man is familiar with the Doppler effect through the sharp drop in frequency (pitch) in the whistle of a train as it speeds by at a crossing. To a passenger on the train, however, there is no change in the pitch of the whistle. The perception of frequency is determined by the number of sound waves received per second. As the train approaches the person waiting at the crossing, there is an increase in the number of sound waves perceived which depends

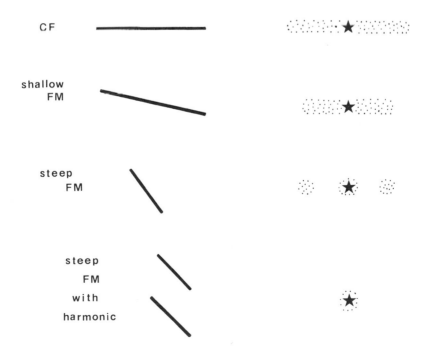

upon the speed of the train. As the train moves away, the number of sound waves per second decreases and the pitch lowers. The Doppler effect is most evident to us at the crossing where there is a dramatic change from higher to lower frequency as the train rushes by. The passenger, on the other hand, is moving at the same speed as the train and the number of sound waves reaching him remains constant. If the train's whistle covered a broader range of frequencies, its effect on the observer would be less dramatic, and by using FM calls that sweep through a range of frequencies, bats achieve some protection from errors associated with Doppler effect.

It would seem logical then that bats whose echolocation calls include pure tone components should be more vulnerable to errors caused by the Doppler effect. However, some species, for example horseshoe bats, exploit the Doppler shifts in frequency associated with wing-beats of insects to detect, and perhaps identify, their targets. The auditory systems of bats such as Old World leaf-nosed species, horseshoe bats, and Parnell's Moustached Bat are highly specialized, and are extremely sensitive to the constant frequency (pure tone) component of their echolocation calls. In the Greater Horseshoe Bat, the specializations permit the bat to measure frequency changes as slight as 10 hertz.

Bats which exhibit these specializations, apparently for detecting flying targets, ensure that the Doppler-shifted returning echoes are in the range of maximum frequency sensitivity by adjusting the frequency of their echolocation calls. Thus, a Greater Horseshoe Bat flying toward a target drops the pitch of the constant frequency component of its call so the Doppler-shifted echo is at 83 kilohertz, its zone of maximum hearing sensitivity. These bats are known as 'Doppler shift compensators,' and their specializations have been compared to those found in the eyes of some birds of prey. Some hawks have two concentrations of rods and cones (receptor cells) on specific points of their retinas that provide a high degree of resolution of the images focused on these points. These concentrations of rods and cones are called 'foveas,' and the concentrations of receptor cells in the inner ears of Doppler-shift-compensating bats have been called 'acoustic foveas.'

For echolocation to work, each outgoing call must be registered in the brain so that the echo can be compared to the original sound. The difference between the two provides the bat with its information, and involves an overwhelming array of neural operations. The muscles of the middle ear and some neural components in the inner ear dampen the loud outgoing pulse to keep it from deafening the bat to the faint returning echo.

Experiments designed to jam echolocation have been relatively unsuccessful. When the frequencies of its calls are blanketed by white noise, the bat tries to keep its back to the sound source, slows down its flight speed, increases the intensity of its calls, and carries on.

Ultrasonic sound, the feature of bat echolocation that foiled early attempts to unlock the secrets of their orientation, is also responsible for one of the principal drawbacks of echolocation. High-frequency sounds are much more vulnerable to atmospheric attenuation or absorption than sounds of lower frequency. The booming bass of a foghorn illustrates how low-frequency sound can carry for considerable distance. To our ears, high-pitched sounds are rarely over 10 kilohertz; most of our conversation is below 5 kilohertz as are the songs of most birds. High-frequency bat sounds are over 200 kilohertz and the echolocation calls of many species range to 100 kilohertz. For calls over 20 kilohertz, atmospheric attentuation increasingly limits the operational range of echolocation in air, probably to about 15 metres in exceptional bats. However, our understanding of the range at which bats detect targets is incomplete, biologists having usually relied on changes in a bat's behaviour to tell them when a target has been discovered. For Little Brown Bats, this reaction distance is usually about one metre, but the bat has probably detected the target well before that. Very recent studies of Big Brown Bats shows that they are able to first detect a 19-millimetre-diameter sphere at five metres.

While high-frequency sound means high atmospheric attenuation and less effective range, it also carries a shorter wavelength and provides the bat with more details about its target. This places the echolocating bat into a cleft stick, if it increases frequency to gain more information about the target, it sacrifices range. No wonder most echolocating bats include an FM sweep in their calls.

Higher-frequency sounds also have a smaller spread, giving a more concentrated beam of sound, and potentially narrowing the area from which the bat gathers information. Bats using high frequency sounds probably compensate for this by moving their heads.

Biologists and naturalists have been aware of bats' abilities of echolocation for more than 30 years, but they usually associate them with ultrasonic sound and do not expect to hear bats with their own, unaided ears. In the course of my studies of bat calls, I have spent many hours listening to these signals when the tape recorder is slowed down from its original recording speed to make the frequencies audible to me. (For example, if one had recorded the calls of a Little Brown Bat at 76 centimetres per second and the calls swept from 100 to 40 kilohertz, slowing the tape eight times would cause the calls to sweep from 12.5 to 5 kilohertz.) Although I knew that Griffin had reported an 'audible bat' from South America, I did not listen for bat echolocation calls when I was out at night. Then one evening in June, 1977, while I was working at a field station in Zimbabwe, a bat caught an insect right over my head and I *heard* its echolocation calls without a bat detector. I later learned that the bat, Martienssen's Free-tailed Bat, was one of several species in the area whose echolocation calls were entirely audible to my unaided ears.

The echolocation calls of this Spotted Bat (15 grams) are readily audible to the unaided human ear. The huge ears and tragus of this species are pinkish in colour, but the three white spots on its back account for its common name and make it one of the most distinctive of North American mammals.

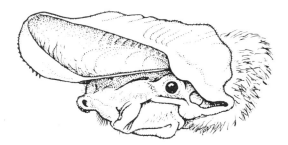

This species, Martienssen's Free-tailed Bat (40 grams), is common in some parts of Africa and is one of the largest and most spectacular of the Free-tailed Bats. This high-flying species produces echolocation calls which are clearly audible to humans. Sketch by Paul Geraghty.

Two years later in the Okanagan Valley of British Columbia, we found another 'audible bat.' The Spotted Bat hunts insects in the Ponderosa pine forests, but many of the local naturalists and schoolchildren had long presumed the familiar calls to be those of an insect. Spotted Bats produce echolocation signals that sweep from 15 to 9 kilohertz, a fully audible range for most people. Had Spallanzani studied one of the audible species, we might have found out about echolocation 250 years sooner.

Humans are not the only animals to use bat detectors. Many insects, notably moths and lacewings, have ears sensitive to ultrasonic sound. The ears may be found on the chests or abdomens of some moths, on the mouthparts of others, and on the wings of lacewings. Moths and lacewings equipped with bat detectors have 40 per cent less chance of being caught by bats than deaf insects.

The moth's anti-bat strategy was largely unravelled by the late Kenneth D. Roeder, an English physiologist who did most of his work in the United States. Roeder demonstrated that the moth plots its defence according to the proximity of the bat. The moth's ear encodes information about the distance to the bat according to the intensity of the bat's calls. A faint call from a distant bat prompts the moth to turn and fly away, and since the average moth hears the average bat at a range of at least 30 metres, it is well out of the way before the bat can detect it. A more intense call from a closer bat prompts the moth to fold its wings and dive to the ground.

Some tiger moths have taken this strategy one step further, for on their chests are noisemakers that seem to jam the bat's echolocation. The jamming is subtle and works by interfering with the bat's ability to process information from its echoes rather than by masking the echoes. Tiger moths equipped with noise-makers only use them as a last-ditch measure, producing their clicks just before they dive to the ground. Recent work by the Canadian biologist James Fullard suggests that bats may have developed a counter to moth hearing, namely using higher-frequency sounds for echolocation.

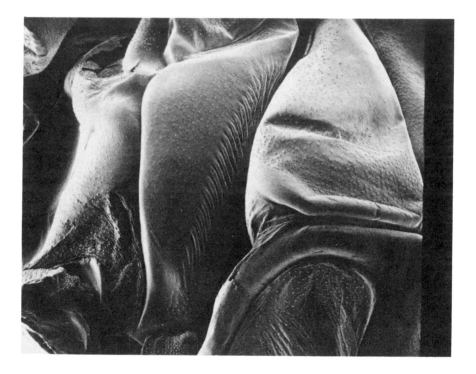

This scanning electron micrograph shows the noisemaker of a tiger moth, apparently used to produce sounds which jam the echolocation of an attacking bat. The noisemaker, which is about one millimetre long, is located on the moth's chest. The series of grooves is responsible for the clicks of the moth.

This Indian False Vampire Bat (50 grams) is typical of the False Vampire Bats in its appetite for a variety of animals. In the wild these bats eat insects, fish, frogs, and other bats, and in the laboratory they listen to sounds of mouse footfalls to locate the mice which they attack, kill, and eat. Note the bifurcate tragus and prominent nose-leaf.

Although temperate bats are relatively conservative in their diets, consuming mainly insects, bats in other parts of the world hunt larger prey. Indian False Vampire Bats can be trained to hunt mice in the laboratory. Because mice can often hear high-frequency sounds, False Vampires usually stalk them quietly, listening to their sounds as they move along the ground. The German biologist Joachim Fiedler showed that Indian False Vampire Bats were behaving like Barn Owls. By producing echolocation calls they could alert the mouse to their approach in the same way other bats alert moths to their impending attack. In Panama, Merlin Tuttle has demonstrated that Finge-lipped Bats hunt frogs which they find by listening to the mating calls of the males. These bats echolocate as they close in on their targets, perhaps because frogs, like birds and people, do not hear ultrasonic sound.

These experiments remind us that while most bats can echolocate, they do use other sources of information to find their targets. We know that some species listen to the sounds produced by their prey while others, particularly fruit-eaters, rely on smell. It remains to be determined just how much insectivorous bats rely on vision as an aid to finding food (but see page 51).

(Opposite) This spectacular photograph by Merlin Tuttle shows a Fringe-lipped Bat (30 grams) attacking a frog. Tuttle has demonstrated that these New World Leaf-nosed Bats locate frogs by listening to their songs, although at this stage of its attack the bat is producing echolocation calls.

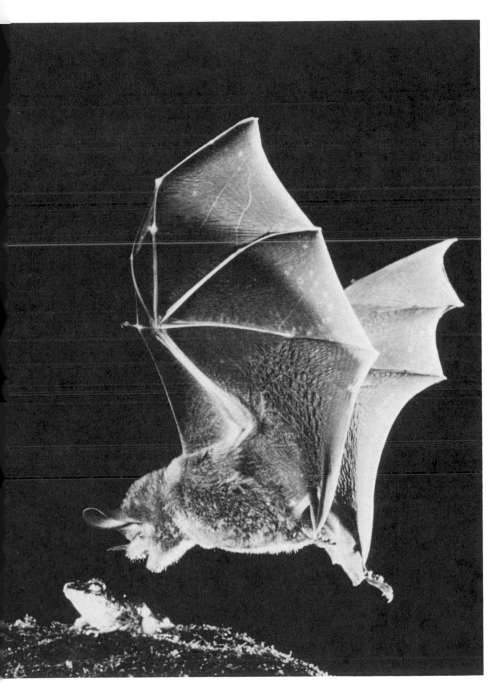

Seeing and Smelling

I can demonstrate that the old saying 'blind as a bat' is misleading by recounting an experience I had while monitoring the echolocation calls of an Egyptian Slit-faced Bat in a small laboratory in Zimbabwe. These bats produce low-intensity echolocation signals, and trying to keep the flying animal within the 20 centimetre range of a microphone forced me to move about continually, presenting the bat with an insistent obstacle to avoid. For the first five minutes after its release in the laboratory, the bat managed to avoid all of the obstacles in the room, including the microphone I kept thrusting in its face. Then, without warning, it turned and flew into the glass doors of a cabinet set against one of the laboratory walls. Reflected in the doors were the bushes and trees growing outside the building. The bat had evidently switched from echolocation to vision to provide direction; although echolocation had provided the bat with a true picture of the glass doors, vision offered a deceptive one.

Like the Egyptian Slit-faced Bat, all of the more than 850 species of bats known to science can see. All bats are believed to have eyes adapted for poor lighting conditions, although in some species, especially those without the ability to echolocate, they are larger and more prominent. So, while the eyes of a slit-faced bat are relatively inconspicuous, those of a flying fox are obvious.

The bat's vision of his world is quite different from ours. Because the retinas of their eyes may lack cones, the light-sensitive cells responsible for colour vision, bats see only in black and white. Furthermore, with the exception of the flying foxes and their relatives, bats lack the tapetum lucidum, the reflective layer under the retina which increases visual sensitivity by reflecting more light on the retina. The absence of this layer can be demonstrated by shining a bright spotlight into the eyes of bats at night. In this situation, you would see none of the 'eyeshine' that would reflect from the eyes of people, cats, some other mammals, birds, and some fish. The tapetum lucidum is prominent among nocturnal animals or those living under conditions of low light. Presumably the layer has not developed in most bats because they have evolved to rely on echolocation to a large degree.

The eyes of this Pallid Bat (20 grams) are particularly prominent. This North American species often flies low over the ground in search of prey.

Experiments with Little Brown Bats have provided some important clues about the vision of echolocating species. Under very dim light, these bats used their eyes to detect obstacles two millimetres in diameter. But, under what we consider normal room illumination, their visual acuity was impaired. It follows, then, that the eyes of Little Brown Bats are adapted for dim lighting, a sensible arrangement for nocturnal animals. However, not all species occupy the dark roosts preferred by Little Brown Bats, and species such as Red or Hoary Bats that roost in foliage may have eyes that function better in brighter light.

Studies of pattern discrimination by echolocating bats also show these species can visually distinguish differences in the shape of objects. In one set of experiments, for example, bats were able to recognize a square, determine that it was different from a similar-sized triangle, and later identify the square when it was presented to them at a different angle.

Radio-tagging experiments with echolocating bats (see page 100) show that vision plays an important role in long-range orientation. By placing radio-tags on Spear-nosed Bats and dividing them into three groups, one with blindfolds, the second with transparent goggles, the third unimpaired, researchers found that the blindfolded group took longer to return to their home roosts than either of the groups whose vision was unimpeded.

Although it is clear that vision plays an important role in the lives of bats, even for echolocating species, more research is needed to clarify its full significance. It would be interesting to know how far they can see, and why certain species use vision in some circumstances and echolocation in others. More work is needed also to determine the degree of bats' binocular vision, although the location of the eyes of some species, particularly some New World leaf-nosed bats, seems to indicate that they have considerable overlap of the images perceived by the two eyes.

OLFACTION

A common objection to sharing space with a colony of bats is the odour from their droppings. Obviously, most bats do not have the same complaint. Many species living in crowded colonies are able to tolerate high concentrations of ammonia (produced from the breakdown of urine) and regularly roost in places that are impenetrable to people not protected by gas masks. In some abandoned white clay mines in Tanzania, tens of thousands of Triple Leaf-nosed Bats roost in dead-end passages where ammonia concentrations are high and there is not enough oxygen to sustain the flame of a pressure lamp. I once visited a colony of Common Vampire Bats in an abandoned gold mine in Colombia where the odour of ammonia immediately brought tears to my eyes and made breathing difficult. Needless to say, my stay in there was a short one.

This insect-eating Northern Long-eared Bat (7 grams) shows the largish ears and relatively large eyes typical of bats thought to locate their food on surfaces; gleaners.

Although their tolerance of near-toxic conditions might lead to the assumption that bats' sense of smell is not particularly sensitive, it appears that odours play important roles in several aspects of their lives. Bats boast an impressive array of glands, the secretions of which probably serve a variety of functions. These glands may be found on the wings, the throat, the shoulder, between the ears, or on either side of the muzzle. Some sheath-tailed bats use the secretions of their wing glands to mark their living space, and some male free-tailed bats use chest gland secretions to mark their females. The pararhinal glands located on each side of the muzzles of Big and Little Brown Bats vary in size seasonally, and when enlarged, produce a clear secretion. Social interactions among Little Brown Bats often involve nose-to-nose contact, suggesting that the secretions of these glands may have some role in their domestic lives. Body odour may also be significant to the interactions between mother bats and their young.

Although many species have distinctive body odours, the role of odour in interactions between species is not clear. These distinctions are often easy to detect: I once found a colony of Bulldog Bats in Puerto Rico because I recognized their distinctive odour in a cave I was visiting there. In other cases, the differences in body odour are more subtle, although some of my students have found they can easily distinguish Little Brown, Northern Long-eared, and Indiana Bats by their odours in the autumn.

Odour also appears to play a part in feeding. Species with a diet of fruit or nectar locate flowers and ripe fruit by smell, but its importance to bats that feed on animals is less certain. Common Vampire Bats have been shown to feed selectively on cows in heat, which suggests they have some means of recognizing them by smell in the same way as bulls. Biologists studying Linnaeus' False Vampire Bats in Panama suspected they used smell to locate the nests of the birds they captured.

Insect-eating bats in captivity usually reject, often with great indignity, offers of insects with bad odours or foul tastes. In one set of experiments, Big Brown Bats tried to avoid eating the sections of carrion beetles that contained glands producing an offensive secretion. If these bats were given only five or six beetles, they ate the entire insect; when they were offered eight or nine, they carefully removed the parts with the offending glands and ate the rest.

Diet

One of the most dramatic measures of the diversity of bats is the variety of food they consume. Although some 600 species eat insects as the main dietary staple, others live on fruit, nectar and pollen, fish, frogs, birds, small mammals, blood, and even other bats. Most of this diversity occurs in the tropics, although many bats from temperate regions vary their diets by eating a wide range of insects.

What any bat eats is determined by two important factors: the need for enough energy to keep the body going, and the need for essential chemicals to maintain it. Just as an automobile requires fuel to move and lubricants to keep the engine running smoothly, animals require protein, carbohydrates or fat for energy, and vitamins and minerals to stay healthy. Since it is essential for bats to consume enough calories to make flight, reproduction, growth, and other bodily functions possible, it is easy to see why such variety is necessary in their diets. Species that feed mainly on plant products require protein which is obtained by eating animals (insects), pollen, or lots of fruit. Bats feeding on animals appear to obtain a balanced diet from this source of nutrition.

To obtain the energy and chemicals they need, bats consume vast quantities of food. Small insectivorous species eat at least 30 per cent of their body weight each night they are active, and in nursing mothers the amount may exceed 50 per cent. These figures appear to apply to all bats, regardless of diet. It may not be impressive to learn that a Little Brown Bat eats three grams of insects on a summer night, but to find 150 mosquitoes in its stomach certainly is, especially when you realize they are not an entire night's ration and the bat probably caught them in less than 15 minutes. The most impressive statistic of insect consumption comes from Texas, where it is estimated that a local population of Mexican Free-tailed Bats eats slightly more than 6,000 tonnes of insects each summer.

Most bats drink water, usually by flying low and dipping their mouths into the surface of a lake or stream. Large concentrations of several species of New World leaf-nosed bats have been sighted at springs in South America. These springs are also frequented by tapirs and other mammals, and it is possible that these visitors obtain some essential nutrient from the water. Some bats live in desert areas and never drink, relying on insect prey for their water requirements.

The basic design of a bat – whether or not it echolocates, for instance – imposes certain restrictions on the range of food available to it. Because they can echolocate, New World leaf-nosed bats that feed mainly on fruit are able to catch insects to supplement their diet with protein. Flying foxes and their relatives, however, do not echolocate and must obtain their protein from something other than insects. Recent studies in the Ivory Coast by the Canadian biologist Donald Thomas suggest that several species of pteropodids get their protein from the fruit that comprises the main part of their diet. The levels of protein in the fruit are low, but this is countered by the consumption of large quantities, and by enzymes in the digestive tract that efficiently extract what little protein is available. Size is also a factor in determining the diet of a bat; a 3-gram Butterfly Bat has fewer prey from which to choose than a 40-gram Large Slit-faced Bat. As a rule, smaller bats are almost entirely insectivorous, and larger species include larger prey in their diet, readily switching to small vertebrates such as fish, birds, and frogs. The smallest bats which feed on plant material are nectar- and pollen-feeders. Larger species more often feed on fruit, but may also supplement their diet with nectar and pollen.

The food selected by bats also depends on where they feed. Flying bats chasing flying insects will not catch scorpions that do not fly, but they often snatch spiders ballooning on pieces of web. Bats feeding on stationary or terrestrial prey often catch resting insects that are able to fly. Bats concentrating their feeding activity over water catch more aquatic insects than those feeding high over the forest, and species hunting over water have opportunities to catch fish not available to high-flying species that visit water only to drink.

There is some danger therefore in placing labels on bats according to their diets. Nevertheless, the most effective way to review the range of food they consume is by examining each food type. Before starting, though, it is useful to remember that the diets of many bats are determined by what is available from season to season, or even from night to night.

Some species eat a wide range of food no matter what the season. In late October 1979, I spent several nights in Zimbabwe's Mana Pools National Park studying the diet of Large Slit-faced Bats with Donald Thomas. These bats carry their prey back to their roosts to consume it, dropping the inedible and less palatable portions to the floor. Although we were unable to make extensive observations because of the country's then unstable political situation, analyses of the discarded food and the layer of droppings in the midden produced an impressive picture of variety. Included on the menu were three species of birds, three species of bats, several species of fish, a frog, dung beetles, hawk moths, giant crickets, katydids, and caddis flies.

The Large Slit-faced Bat (35 grams) from Africa is an example of a bat-eating bat, taking individuals up to 10 grams as they either roost or fly about. The facial ornamentation is typical of the Slit-faced Bats.

Further studies of Large Slit-faced Bats are needed to clarify how different prey are captured and how the diets of individual bats vary. Studies of this species are also needed to produce more evidence about the flexibility of the hunting habits of bats in general. Keeping this flexibility in mind, let's begin our dietary tour of the bat world, starting with bats that feed on other animals.

ANIMAL PROTEIN

About 700 species of bats, all of them from the suborder Microchiroptera, and all with the ability to echolocate, eat animals as the main part of their diet. There have been occasional records suggesting that flying foxes and their relatives, members of the Megachiroptera, may also catch and eat animals, but hard evidence of this is lacking and must be presented before these apparent vegetarians can be safely included in this category.

Animal-eating bats all have similar tooth design: the incisors or front teeth tend to be small, the canines or eye teeth long and prominent. Some premolars, the teeth between the canines and the molars, are small and often absent, while the premolars closest to the molars are often large. Cheek teeth (molars) feature V- or W-shaped cusps designed to shred prey rapidly through an effective combination of cutting and crushing. Unlike wolves, weasels, and cats, bats do not have carnassial teeth specialized for cutting muscle and sinew. As a result, bats that feed on small mammals have difficulty chewing them up effectively, and often spend several minutes munching each morsel. For example, I have watched a Large Slit-faced Bat catch and kill a small Noack's Leaf-nosed Bat, an operation which took only a couple of minutes. However, eating this meal took the bat almost five hours!

Animal-eating bats use a variety of strategies to capture their prey. Bats seeking flying or non-flying prey may fly continuously as they hunt, or they may sit on a perch waiting for a target to pass within range, taking off only to make the capture. Although some bats use only one of these strategies, others, such as Heart-nosed Bats of east Africa, may use both.

Even if their prey is not airborne, most species of insectivorous bats are thought to fly continuously while hunting. Fish-eating bats that have been studied also fly continuously back and forth over lakes and streams, taking food as the opportunity presents itself. Insect-eating bats can be further divided into two categories, species detecting targets at long range (more than one or two metres), and those operating at shorter range (less than one or two metres). Long-range bats usually fly high and fast and make one capture attempt during each pass through a swarm of insects; short-range species hunt in more confined areas and often make several attempts at capture on one pass through a patch of prey.

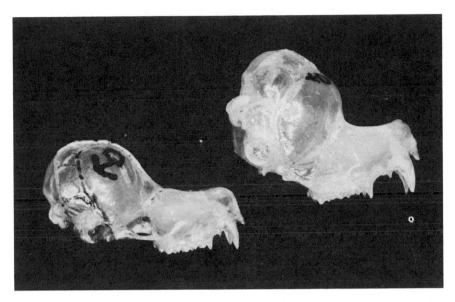

Differences in the shape of the skull translate into striking differences in the faces of bats. Compared here are the skulls of a Parnell's Moustache Bat (lower left) and a Blainville's Leaf-chinned Bat.

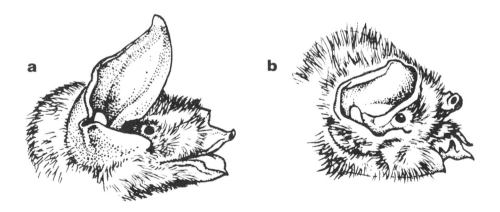

The corresponding faces show the two species (a, Parnell's Moustache Bat; b, Blainville's Leaf-chinned Bat) with flesh and fur and ears. Both are insectivorous and occur in the Neotropics. Sketch by Paul Geraghty.

A California Leaf-nosed Bat (8 grams) which locates prey by echolocation, vision, or sounds emanating from the prey itself. The prominent nose-leaf is typical in design of New World Leaf-nosed Bats. This species is insectivorous.

While continuous flying may increase the amount of prey a bat encounters, it also requires more energy. As a result, some bats have adopted several techniques to capture as much food as possible in the shortest flight time. In the tropics, for example, it is common for millions of winged ants or termites to emerge at one time. Bats quickly detect these rich patches of food, sometimes by eavesdropping on the echolocation calls of other bats. Furthermore, insectivorous bats are fast eaters. For example, Little Brown Bats chew their food seven times per second. Several insect-eating bats have also developed cheek pouches designed to hold food for careful chewing later.

Other insect-eaters minimize the energy used in flying by waiting for their prey to come to them. Heart-nosed Bats sometimes scan the ground from perches about one metre up, attacking prey such as crickets, scorpions, centipedes, and millipedes they detect as it moves around. Giant Leaf-nosed Bats wait on perches for insects to fly by, selectively attacking prey with direct flight paths and ignoring those flying erratically. A straight flight path makes the position of the target predictable, an asset to the bat in plotting its flight path for interception.

Although echolocation is employed in most of these strategies, some bats that hunt prey moving on the ground use other cues. Indian False Vampire Bats often turn off their echolocation when attacking mice, relying on the rodents' footfalls to locate their targets. Fringe-lipped Bats rely both on the calls of frogs and on their own echolocation calls to find their prey, and Gary Bell, a Canadian biologist working at Carleton University, has recently found that California Leaf-nosed Bats use their eyes to find food on the ground.

Although the amount of research on the hunting behaviour of bats has blossomed in recent years, the strategies of most bats remain a mystery. Only more observation and experimentation will allow us to add to the list of hunting techniques with which we are already familiar.

Insects

The diets of insectivorous bats have been documented in several ways. Insects can be identified by analysing the stomach contents and droppings of the bats, or by looking for discarded pieces of prey at roosts. There are also some observations of bats feeding on particular insects.

Although many of the more than one million species of insects in the world are invulnerable to bats, the list of those included in the collective diet of bats is long and impressive. Bats will eat moths, grasshoppers, large insects like dung beetles (eight grams), and small ones like caddisflies (100 milligrams), insects that fly, and those that are flightless. To locate insects bats will swoop low over the surface of lakes, snap them out of the air, and even land on the ground and pursue them 'on foot.' Although some insect-eating bats grab their victims directly in their mouths, others scoop them out of the air with their wings or the membranes enclosing their tails.

Variety truly is the spice of life for insectivorous bats, whose menu of prey may change from night to night or season to season. In one evening, a Little Brown Bat may feed exclusively on mayflies, the next night on midges, and the third night on a mixture of mayflies, midges, moths, and beetles, depending upon what is available where it hunts. Some African species appear to focus on moths during the wet season when these prey are abundant, but eat a variety of insects in the cold dry season when insects are scarce.

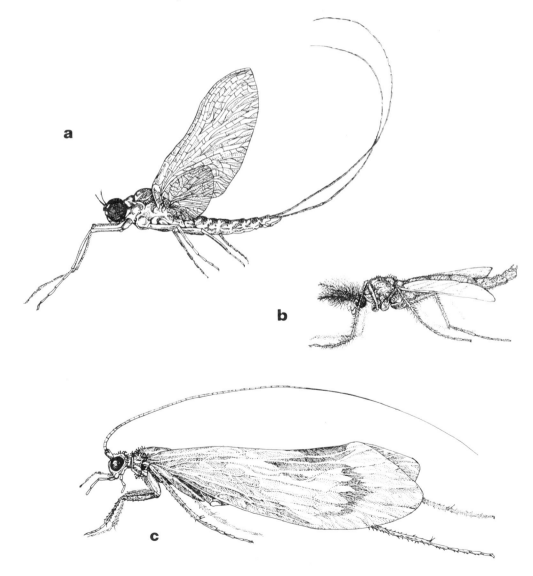

Insect-eating bats consume a wide range of prey, including many different kinds, sizes, and shapes of insects. Aquatic insects such as mayflies (a), midges (b), and caddisflies (c) are the stable food of some bats, while others feed on dung beetles (d), mosquitoes (e), moths (f), crickets (g), or katydids (h). Sketches by Connie L. Gaudet.

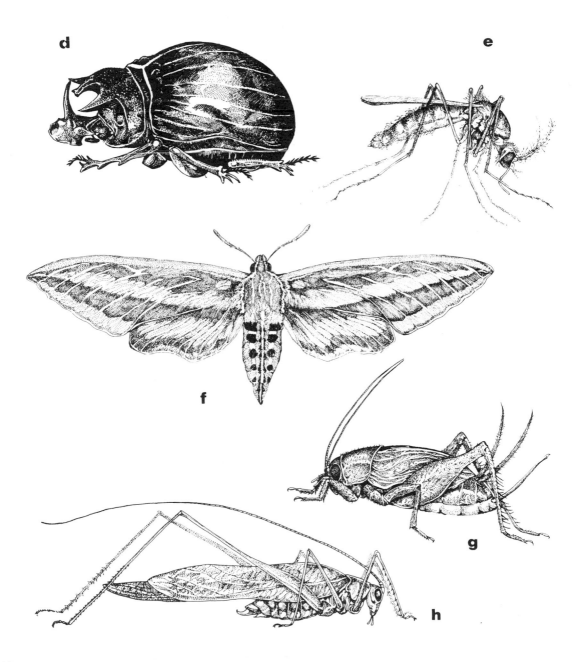

There is little evidence that any species of bat specializes on a particular type of insect, or even a group of insects such as moths, beetles, or flies. Moths with ears sensitive to the ultrasonic echolocation calls of bats do not appear to hear all bats equally well. This has led to the suggestion that some bats, perhaps because their very high-frequency echolocation calls are undetectable by moths, are more adept at catching moths than others. Although the picture is still incomplete, this theory is strengthened by the African Trident-nosed Bat, which feeds almost exclusively on moths and has an echolocation call above 200 kilohertz. However, we know little about the hunting behaviour of this bat, a missing piece in the puzzle.

Fish

About six species of bats regularly eat fish, but most is known about the Mexican Bulldog Bats of South and Central America. Using echolocation to detect ripples created by the fins or heads of fish swimming at the surface, these bats swoop down and gaff their prey with the sharp claws of their enlarged hind feet. Several other species, including Lesser Bulldog Bats and Mexican Fishing Bats, also gaff fish, and this method may be used by four or five species of Mouse-eared Bats in different parts of the world. It is not known how Large Slit-faced Bats or Indian False Vampire Bats catch fish.

Mexican Bulldog Bats and Mexican Fishing Bats do not restrict their diets to fish. Analyses of their stomach contents and feces indicate they also eat a variety of insects, mainly species frequenting water surfaces, sand bars, or beaches. Like fish swimming at the surface, these insects are conspicuous to a bat hunting over or near water; they are also captured in the same manner, with the claws of the hind feet. A colony of Mexican Bulldog Bats I visited in Puerto Rico fed largely on millipedes, but I did not discover where or how they were captured.

(Opposite) Carl Brandon took this excellent photograph of a male Mexican Bulldog Bat (60 grams) snatching a grasshopper from the surface of the water. These and other bats often feed on fish swimming close enough to the surface for their heads, tails, or dorsal fins to protrude, providing the bats with a target they can locate by echolocation.

Large Slit-faced Bats often use their
wings to support their food, particularly
when the prey item, like this frog, is
relatively large. These bats, which occur
in parts of Africa, typically hang by one
foot while feeding. Sketch by Connie L.
Gaudet.

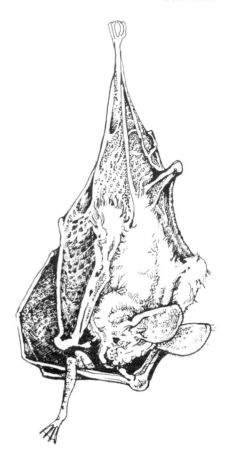

Other Vertebrates

Although it is not known whether any species of bat feeds exclusively on specific
vertebrates, we do know of some that eat a variety of animals with backbones.
Large Slit-faced Bats include fish, birds, frogs, and other bats in their diet.
Linnaeus' False Vampire Bats from Central America will eat birds, sometimes
concentrating on Groove-billed Anis they catch in their nests, but also feed on
mice and bats. If we are to use these as examples, it is probably safe to say many
larger bats (over 40 grams) that hunt non-flying prey regularly eat small
vertebrates and a range of other animals.

Carnivorous bats usually attack their victims by going for the head, the
quickest way for any predator to subdue its prey. Where they differ among
themselves is in the means of launching the attack. Fringe-lipped Bats from the
Neotropics commonly hover over frogs, capturing them by swooping down and

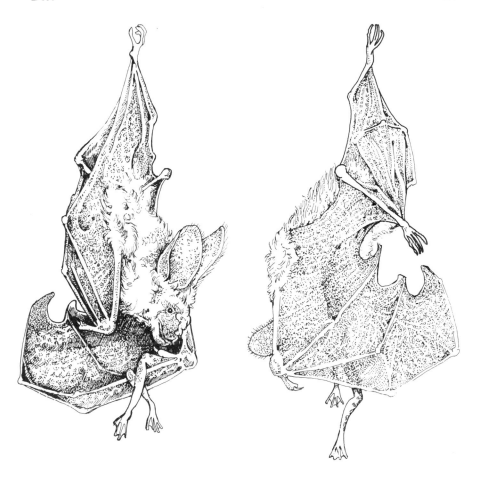

grabbing them with their mouths. Others, including Indian and Linneaus' False Vampire Bats, more often tackle their prey by landing on top of them and biting their heads. Large Slit-faced Bats will fling themselves to the ground over a frog or large insect, or hurl themselves against a wall to catch a roosting bat. In other cases these voracious hunters catch flying bats in mid-air by grabbing them in their mouths, again aiming their attack at the head.

Biologists are only beginning to understand how often 'cannibalistic' bats locate other bats by eavesdropping on their echolocation calls. Recent studies in Zimbabwe suggest that Large Slit-faced Bats do not rely on these signals, although they did respond to 'irritation squawks' and other social vocalizations. What the studies do show, however, is that the movement of prey is crucial to detection and capture. More studies are needed in this area, though, and I look forward to the results of experiments that will clarify the importance of eaves-dropping to bat-eating bats.

Blood

Only three species of bats, all from Central and South America, feed exclusively on the blood of living animals. Although the natives were familiar with them long before the arrival of Europeans in the sixteenth century, our name for them comes from the Magyar word 'vampir,' meaning a person who comes back from the dead to feed on the blood of the living. Although many other bats from the Neotropics have scientific names that reflect man's fascination with blood-eaters (for example, *Vampyrum*, *Vampyrodes*, *Vampyrops*, and *Vampyressa*), none of these actually feeds on blood. Oddly, the true vampires, named *Desmodus*, *Diaemus*, and *Diphylla*, do not have scientific names that include the 'vampir' root.

Unlike the horror-movie presentations of vampire bats as huge animals, these species are actually medium-sized. The adults weigh about 40 grams and have wing-spans of about 50 centimetres. Movie-makers usually cast flying foxes in the role of vampires, and use the nightly exodus of Mexican Free-tailed Bats from locations like Carlsbad Caverns to depict swarms of vampires. A Mexican Free-tailed Bat was used as the vampire in the movie *Papillon.*

Vampires use their razor-sharp incisor teeth to make a quick gash in the skin of their victim, having used their cheek teeth as scissors to shear away intervening fur or feathers. Their saliva contains an anticoagulant to keep the blood from clotting. Once it has made the incision, the vampire drinks its fill, usually about 10 cubic centimetres, and flies away, leaving only its distinctive mark, a shallow cut surrounded by a smear of dried blood. Vampire bats prefer to attack sleeping victims, usually cattle and other domestic animals; they only occasionally attack man. As a rule, the victim does not wake up during the attack.

Biologists are most familiar with Common Vampire Bats, which prefer to feed on the blood of mammals. Hairy-legged Vampires prey mainly on birds, and the diet of White-winged Vampires includes blood from both birds and mammals. Vampires usually feed on parts of the body with little fur or few feathers. When they prey on people, parts of the body protruding from under the bedding – cheeks, forearms, nose, and toes – are usually attacked. On cattle, donkeys, and horses, vampire incisions are most often made in the neck and shoulder area, the base of the ears, on the back above the hips, or around the hooves. There is some evidence that vampires from different areas prefer to attack different parts of their victims' bodies, although the reasons for this are unknown. A study by Dennis Turner, as a student at Johns Hopkins University in Baltimore, demonstrated that the sites chosen for attack by vampires changed seasonally with the number of cattle available.

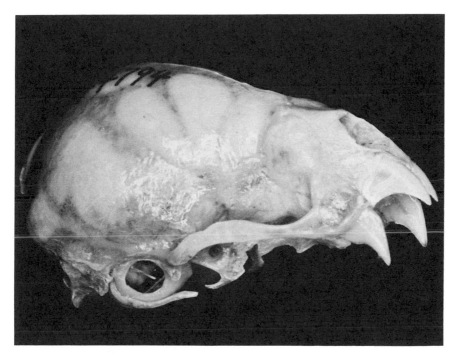

The razor-sharp incisor (front) teeth of this Common Vampire Bat (skull 23 millimetres long) are used to make a shallow cut in the skin of a victim. The prominent canine (eye) teeth are used more often in squabbles between bats, while the small teeth behind the canines serve to shave fur or feathers from potential bite sites.

Vampire bats are beautifully adapted for their diets. Because blood is heavy and relatively difficult to digest, Common Vampire Bats begin to urinate within seconds of starting to feed. The urine contains mainly the non-nutritive elements of the blood meal, the electrolytes in the blood plasma. This strategy allows the bat to rid itself of useless weight, while increasing its capacity for food.

Vampires are also well adapted for taking off after consuming a heavy load of blood. Common Vampire Bats crouch low and fling themselves into the air, using their extra long thumbs as throwing sticks to gain leverage. There are records of Common Vampires eating too much blood to allow takeoff, but they counter this problem to some degree by exceptional agility on the ground. Vampires often show this prowess by landing beside and walking up to their sleeping victims.

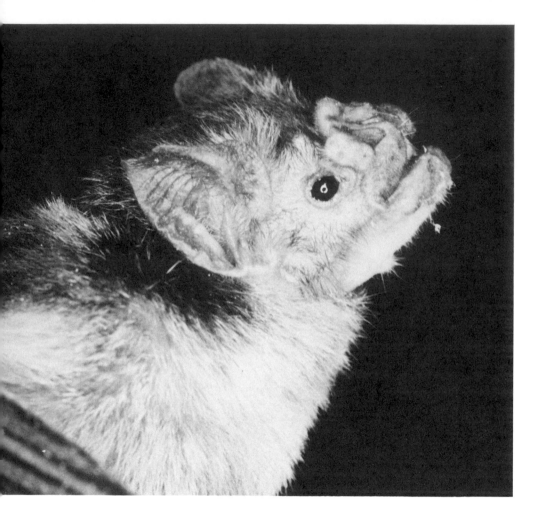

This Common Vampire Bat (60 grams) lives in the Cornell University Colony.
Although vampires are usually placed in the New World Leaf-nosed Bats
(Phyllostomatidae), the picture makes it clear that the nose-leaf is much reduced over
that of other species in this family (compare the California Leaf-nosed Bat and
Fringe-lipped Bat, pages 50 and 39).

Although vampires have well-developed senses of smell and vision, they rely in part on echolocation to detect their prey. Their calls are low in intensity, though, and detectable only when they fly close to a sensitive microphone. It is possible that the intensity of their calls is related to the hearing abilities of potential prey. It may be significant, for instance, that dogs, known for their ability to hear high-frequency sounds, are rarely the victims of vampires. There is no conclusive evidence for this, but it seems logical that vampire bats producing very intense echolocation calls would have trouble sneaking up on a victim that could hear their signals. It has recently been reported by workers in Germany that Common Vampire Bats have pits located on their faces which are sensitive to heat. These structures probably assist the bats in locating suitable places to bite, and are comparable to the heat-sensitive pits of rattlesnakes.

In themselves, the bites of vampire bats are not dangerous. But because vampires may spread rabies (see page 139), their bites carry the potential for danger, and people should avoid feeding them. Rabies spread from vampires to cattle has created a serious economic problem in some areas of South and Central America. The situation has prompted many attempts to control vampires, including the indiscriminate use of hand grenades and napalm to attack them in their roosts. A more sophisticated and workable approach is to apply a paste-like poison to the vampires actually feeding on cattle; the treated bats return to their day roosts and groom one another, spreading the poison around the colony and killing only the guilty bats. Another approach to the problem would be inoculation of the livestock, but that could prove as costly as the labour-intensive trapping and treating of offending bats.

The evolution of vampire bats is largely a mystery. They are closely related to and classified with the New World leaf-nosed bats, although they have been treated as a separate family. It is not known what vampires feeding on the blood of mammals subsisted on before the arrival of Europeans with their domestic animals. People are not the preferred food of these vampires, and in the recent past (10,000 or 20,000 years) there were relatively few large mammals in South and Central America. These bats must originally have relied on a variety of smaller mammals and birds; with the arrival of Europeans and their large numbers of domestic animals, the vampires, reflecting a typical pattern of bat behaviour – opportunism – were quick to take full advantage of this new, accessible, and abundant prey.

Vampires remain the most infamous of bats. People continue to assume that what is true about vampires and rabies is true of all bats. It is true that Common Vampire Bats have adverse effects on the economies of some countries, but vampires also represent a pinnacle of adaptation in the bat world. Their specializations, in teeth, kidneys, and launching ability, are exquisite. It is unfortunate that vampires give other bats a bad name; in other circumstances any bat would do well to be compared to such adaptable creatures.

The relationship between bats and plants involves a high level of co-evolution. The fruit and flowers of plants have been made more accessible to bats in exchange for their help in pollination and the dispersal of seeds. For their part, bats have developed special morphological and behavioural designs to find and feed on plants.

Nectar, pollen, and fruit are the three main plant products eaten regularly by bats. But plant-eaters must also have protein, and they find this in several ways. Some bats stick to diets made up of plant products, but obtain sufficient protein in flowers and large amounts of fruit. Other species eat insects, although it is difficult to determine the amount of insects consumed by fruit- and flower-feeding bats because of the way they are eaten. Short-tailed Fruit Bats, for instance, chew insects and suck out their tissues and fluids, ingesting little of the exoskeleton. Since the exoskeletons are used to identify the remains of an insect meal, analyses of the stomach contents feces of this bat fails to reveal the true proportion (or even the presence) of insect protein in the diet. Only field studies of the behaviour of bats will clarify the contribution of different foods to the diets of fruit- and flower-feeders.

Fruit

Bats feeding on fruit ingest mainly the pulp and juices, chewing and sucking carefully to separate these parts from fibrous materials and other indigestible components. Most bats seem unable to digest seeds; large seeds are usually not swallowed and the smaller ones pass undamaged through the bat. Food passes quickly through these bats, usually in about 20 minutes, reducing the time they have to fly about with full stomachs. Fruit may be eaten right on the plant, but more often it is carried to a night roost. the advantages of this are not clear, but it could be a way to reduce the risk of being nabbed by a predator.

Bats eat a variety of fruit, including important cash crops such as mangoes, papayas, bananas, guavas, and custard apples. In some areas, groups of fruit bats, particularly flying foxes, can demolish a crop of ripe fruit. What fruit is eaten by bats depends on what is available, and usually this changes with the seasons. In Africa, tropical Asia, and Australia, some fruit-eating bats migrate hundreds of kilometres to follow the changing availability of food.

Two patterns of searching for food are well known. Smaller species have wide search areas and quickly locate patches of suitable fruit. Larger species learn the

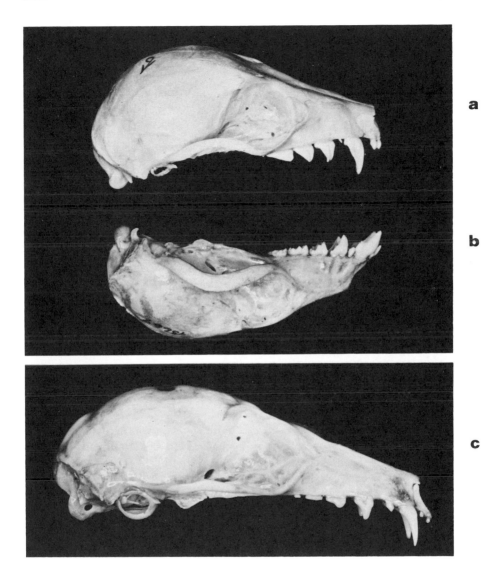

The skulls and teeth of two fruit-eating bats are compared with those of a nectar-feeder. In Taylor's Flying Fox (a), the molar or cheek teeth are squared, while those of the Harpy Fruit Bat (b) have many cusps. Woermann's Long-tongued Fruit Bat (c) has peg-like teeth. The skull of the Harpy Fruit Bat is about 43 millimetres long; that of Woermann's Long-tongued Fruit Bat 26 millimetres in length.

locations of suitable plants and visit them selectively. Experiments in which artificial concentrations of fruit were placed in the field showed that smaller species of bats quickly detected the new patches of food, while the larger bats usually ignored them.

Fruit-eating species have relatively broad and flat teeth for pulping their food. Across the roofs of their mouths are pronounced ridges used like a washboard in conjunction with the tongue which presses the fruit against the ridges while the bat sucks to extract the juices and pulp. Harpy Fruit Bats have molar teeth covered with extra cusps which may serve as a sort of fruit-juicer.

Fruit-eating bats often become covered with the sticky juices from their food and spend considerable time grooming themselves between courses and after dinner. When eating figs, a Streseman's Dog-faced Bat hangs upside down by one foot, using the other to hold the fruit. The bat takes chewed pieces of fig, packing them into the wing membrane running from shoulder to thumb until both of these pouches are filled. What's left is eaten, followed by the fruit packed in the wings. This takes about 30 minutes, and the bat spends another 10 minutes grooming its fur and licking the pouches clean before picking another fig and starting again.

Nectar and Pollen

Many species of bats are highly specialized for feeding on nectar and pollen. Their tongues extend to unusual lengths, often up to one third the length of their bodies. This extension is achieved in part by contracting muscles in the tongue to push blood to the tip, much as the end of a party toy unrolls when you blow into it. The tip of the tongue is covered with conical papillae that increase the surface area and the bats nectar-lapping capacity. Furthermore, the scales on their hairs extend from the shafts and act as pollen traps. The scales on the hairs of bats that do not feed on flowers usually adhere closely to the shafts.

In tune with a diet requiring little chewing, the teeth of nectar-feeding bats are small in size and few in number. Some species of New World leaf-nosed bats lack lower incisors and their cheek teeth have been reduced to small pegs. The snouts of these bats are elongated, giving them the appearance of hummingbirds.

(Opposite) Donna Howell took this photograph of a Sanborn's Long-tongued Bat (17 grams) visiting a cactus flower, from which it obtains nectar for energy and pollen for protein.

Because flower-feeding bats can be important pollinators, some flowers have developed specializations to attract and exploit bats. These flowers release their pollen at night and usually give off a strong odour, often the stale smell of fermentation, and offer large amounts of nectar. Many bat flowers provide easy access to their pollinators by hanging below or extending from the foliage of the plant. They are usually supported by sturdy stalks that will not break under the weight of the bat. An excellent example of a plant specialized to attract bats is the India Trumpet-flower of Malaysia, a 26-metre-high tree visited nightly by one species of Dawn Bats. According to a study by the American zoologist Edwin Gould, each tree produces from one to 40 stalks holding groups of flowers that extend well out from the canopy. Each night of the year, one to four flowers open on every stalk, usually between an hour and a half and four hours after dark. The flowers remain open through most of the night, but are finished and drop to the ground before dawn. The Dawn Bats time their arrival at the trees to coincide with the opening of the flowers. The impact of these 60-gram bats landing on the open flower gives them easy access to the nectar. Lighter bats, or Dawn Bats arriving too early, cannot provide enough force to open the flower and reach the nectar.

These scanning electron micrographs of the hairs of Geoffroy's Tailless Bat (a), Woermann's Long-tongued Fruit Bat (b), and a Mexican Bulldog Bat (c) show that the hairs of different species can be distinctly different. Note that the scales on the shafts of the hairs of Geoffroy's Tailless Bat and Woermann's Long-tongued Fruit Bat protrude more than those of the animal-eating bat. Protruding scales in flower-visiting species have been proposed as pollen traps, permitting the bats to carry pollen more efficiently. The similarity of the hairs of the two flower bats is striking, given that they are from different families (a, New World Leaf-nosed Bats; b, Old World Fruit Bats). The hairs are about a tenth of a millimetre in diameter.

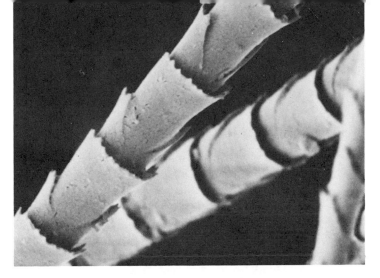

 a

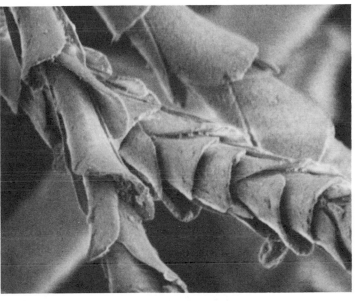 b

 c

The specializations for finding and feeding on nectar and pollen have resulted in a decrease in the acuity of echolocation in New World bats, and probably less reliance on insects as a source of protein. Nectar seems to provide the required carbohydrates, and proteins are obtained from the pollen. Since pollen is very resistant to digestion, some New World nectar-feeders create a milieu for breaking it down by drinking their own urine. The bats in question, Sanborn's Long-tongued Bat, live in Mexico and the southwestern United States.

The ability of bats to exploit a wide range of foods is one of their most fascinating characteristics. The specializations required for different diets offer textbook examples of structure-function relationships in the design of animals, and their interactions with food, from hearing defences of insects to co-evolution with plants, provide a showcase of evolution.

Energy and Survival

Where a bat roosts, when at night it is active, how it responds to rain and cool weather, all involve its energy demands. The bat flying about early on a summer evening and the bat hibernating in a cave in midwinter have each solved important energy problems. The flying bat is dispersing a surplus of heat; the hibernating one is coping with a shortage of heat. Because of different climates, not all bats face the same energy problems.

It seems bats are always facing a problem of too much or too little heat. When they fly, the problem is usually excess heat, produced by the contraction of large flight muscles. Since body temperatures much above 40° C affect vital processes and usually lead to death, it is crucial that the bat find a way to avoid overheating.

To cool off, bats appear to rely on their wings. Heated blood is pumped from the body into the wings where it comes in close contact with the air, and the excess heat dissipates. The evidence for this radiator effect is provided by the presence of two systems of blood vessels in the wings. In one system, the blood passing through arteries into the wings moves into smaller arteries, capillaries, smaller veins, larger veins, and then out of the wings and back to the heart; from the capillaries, the blood supplies the wing tissues with oxygen and nutrients and removes accumulations of wastes. In the second system, the one responsible for the dissipation of heat, the small arteries connect directly to small veins with no intervening capillary bed. Blood passing through the first system undergoes a drastic drop in pressure as it traverses the capillary stage, but there is no such plunge in pressure in the cooling system. Direct connections between arteries and veins are not common among mammals, and bats have developed the system to solve a special problem.

Like many other mammals, including man, bats also use evaporation to avoid the dangers of overheating. When an animal sweats, the water on the skin evaporates, absorbing heat from the body. On very hot days, bats and other mammals often lick themselves to put saliva on their fur and skin to increase the amount of evaporative cooling. Some animals may even urinate on themselves to add to the liquid on their bodies. Bats often use their wings as fans in an effort to increase heat loss through evaporative cooling. This method of cooling has obvious drawbacks – the energy involved in flapping of wings, and the loss of water. A cool bat that dies of dehydration has merely traded one problem for another. Indeed, the ability to avoid heat and conserve water is as important to some species as the ability to lower body temperature is to others.

In flight, bats burn energy at about the same rate as an active shrew or flying hummingbird; in all three animals fuel consumption is reflected by heartbeat rate, over 1300 beats per minute. The ability to sustain flight depends upon available food, either carried as body fat and other chemical energy or as food eaten and digested in flight. Upon landing, a bat's energy consumption and heartbeat rate decline drastically – the latter to about 200 beats per minute – and the animal consumes stored or recently acquired fuel at a much slower rate.

When there is enough food, bats have no problem compensating for the energy burned in flight. Indeed, as we have seen, bats have adopted several strategies that allow them to feed efficiently and minimize the energy used in flight. However, when food is scarce, bats must rely on other ploys. One way to survive long periods without food is to reduce the rate of fuel consumption, by lowering the body temperatures and entering a state of torpor. Torpor over a long period is called hibernation, but bats which hibernate may also use torpor as a means of surviving short inclement periods and conserving energy.

Only plain-nosed and horseshoe families of bats include species that hibernate. A hibernating bat can tolerate lower-than-normal body temperatures for prolonged periods, and then can raise its temperature back to normal without external stimulation. In some species of hibernating plain-nosed and horseshoe bats, body temperatures often drop from 40°C to 5°C. Most other species of bats are not able to survive significant drops in body temperature, although there is evidence that some free-tailed bats can withstand temperature reductions approaching those tolerated by their hibernating cousins.

A distinction must be made between the torpor of hibernation and that of a summer day or night. A hibernating bat is torpid, but a torpid bat is not always hibernating. Bats achieve torpor by selecting cool roosts and allowing their body temperatures to drop to the level of their surroundings. Maintaining their body temperatures below their surroundings would require consumption of energy, just as it takes fuel for a man to air-condition his home. Plain-nosed and horseshoe bats enter torpor readily, reducing their body temperatures and heartbeat rates by prudent selection of cool roosts. At temperatures of 1°C to 5°C, the range commonly encountered in hibernation sites of Little Brown Bats in Canada, these bats would have heartbeat rates of 10 to 15 per minute, evidence of considerable reduction in energy consumption (the same bat at 25°C would have a heartbeat rate of 100 per minute).

The distribution of any species of bat is related to its ability to endure changes in body temperature, the energy efficiency of its diet, and the avilability of food. Plain-nosed and horseshoe bats that feed on insects and tolerate body temperatures from 0°C to just over 40°C are found from the north to the south temperate regions. On the other hand, slit-faced bats and Old World leaf-nosed bats, insectivorous species with little tolerance for cold and no ability to lower body temperature, occur only in the tropics. These bats require a dependable supply of food for flying and heating, and lowered body temperatures are fatal to them. Bats feeding on nectar and pollen and fruit seem more restricted by the availability of food than by temperature tolerance, although flying foxes are not able to adapt to temperature fluctuations as well as most New World leaf-nosed bats, their dietary counterparts in the Americas. However, in this argument it is difficult to separate cause from effect; perhaps if there were more abundant food resources the bats could afford to live in cooler areas.

To appreciate the energy balance of a typical bat, it is useful to use the concept of the 'thermal neutral zone,' a phenomenon common in warm-blooded animals. In summer if I stand quietly in the shade of a tree, there is a range of temperatures that I find comfortable, even if I am clad only in a T-shirt and shorts. Within this range, I do not have to produce heat by shivering or cool off by sweating or fanning myself. This is my thermal neutral zone, the range in which my body uses no energy either to keep warm or to lose heat in order to maintain a normal 37°C temperature. However, if I move into the sunshine or if the outside temperature drops suddenly, I am placed outside the thermal neutral zone. I am then forced to use energy to keep my body temperature constant, an expenditure reflected in changes in my pulse rate.

Furthermore, my thermal neutral zone is not static. If I begin to run, I increase my consumption of energy, reflected by an increasing pulse rate (from, say, 60 to over 120 beats per minute). Burning this energy also produces heat from contraction of muscles; as a result, the air temperature is no longer comfortable and I begin to sweat, taking advantage of evaporative cooling to unload body heat.

Most species of bats have thermal neutral zones in the same range as ours, and must select roosts which are within this zone to minimize energy consumption. Plain-nosed and horseshoe bats, by virtue of their ability to enter torpor, are less restricted in their roosting habits for most of the year.

SUMMER

In temperate regions, bats respond to the thermal challenge in part by exploiting different roosts. Day roosts are used to pass daylight hours, and night roosts serve as refuges for eating and digesting between bouts of foraging. Because some of their needs differ, males and females do not always use the same roosts. For both sexes higher body temperatures mean faster digestion and quicker availability of the energy obtained by feeding. But higher temperatures also mean that energy is used more quickly. the 'ideal' roost, then, depends on its function; the best night roost may be slightly warmer than the thermal neutral zone, while day roosts may be a bit cooler to reduce the rate of energy consumption.

Pregnant or nursing females of many species congregate in day roosts that serve as maternity wards and nurseries. These sites rarely harbour adult males, and their warmer temperatures tend to increase the growth rates of the young. Thus an old barn housing Little Brown Bats is a mosaic of roost conditions. Males roost in cooler positions, usually alone or in small groups, while pregnant females and young gather in large groups in a nursery which is usually located close to the roof. On cool days the females and young cluster together to keep warm, and the males, in their cooler sites, enter torpor. By clustering, each bat minimizes heat loss by exposing less of its body surface to the surrounding air and benefits from contact with a warm neighbour. On warmer days, the females and young may spread out to keep cool, and on very hot days they may move down from the roof areas and occupy lower positions. One study in California showed that the bats adjusted their positions on the roof boards so that they remained in their preferred temperature zone. As the season progresses, the day roosting behaviour of females changes and they abandon the hot, crowded nurseries for cooler roosts where they can begin to conserve energy and build up their condition after bearing and raising young. In temperate areas this change in roosting behaviour is a prelude to migration and/or hibernation.

In parts of the world where the temperatures in caves are too low to make them useful as nursery sites, some species use their collective body heat to raise the temperature of small ceiling chambers and turn these into maternity sites. For this approach to be effective, though, several thousand bats may be needed, and the cave temperatures must not be too low. In places such as Canada, where caves are too cold for this strategy, nurseries are located in other typical day roost settings such as buildings or crevices or hollows in trees.

In Canada male Big Brown and Little Brown Bats frequently roost alone or in small groups on cool days, often entering torpor as their body temperatures fall below the thermal neutral zone. This is an efficient means of conserving energy, and because males are not actively involved in raising their young, it is a luxury they can afford. Interestingly, these males, and females without young, usually select roosts with southern or western exposures, thus obtaining not only afternoon warmth but also a built-in alarm clock to wake them for an evening of activity. As we shall see, bats in summer torpor require an external stimulus to rouse them.

Although winter is a time of cold, snow, and ice in much of the world's temperate zones, winter in the tropics is more often characterized by cool, dry conditions. Winters in both regions often bring drastic reductions in insect populations. Insect-eaters can react in several ways to the food shortage; they can migrate to warmer, moister sites where food will be somewhat more plentiful; they can hibernate; or they can switch to a different source of food. Bats tend to use the first two solutions, and although we tend to associate migrations with birds, many temperate bats use a combination of migration and hibernation to get around local food shortages. Migration is discussed in a later chapter.

The big problem facing hibernating bats is conservation of energy. Hibernating ground squirrels have two stores of food for winter, their body fat and a hoard of food in their burrows. Hibernating bats have only their body fat to see them through winter. It is therefore crucial that they select sites with temperatures that will guarantee that their store of fat lasts the duration. The sites must be cool enough to keep their metabolism rates low, and warm enough to keep them from freezing or burning precious energy to remain unfrozen. For this reason, hibernation in exposed sites such as hollow trees is feasible only in climates where there are no prolonged periods of subfreezing temperatures. Otherwise, hibernation is restricted to underground environments such as caves or abandoned mines. A bat hibernating in a hollow tree and awakened by freezing temperatures may be able to survive only by burning considerable energy, often to find a warmer site. Bats hibernating in a cave, on the other hand, can usually afford to wake up in such circumstances because sites where temperatures are above freezing are often nearby in the same cave. For example, Little Brown Bats that enter hibernation in peripheral parts of a cave or mine will arouse if the temperatures at these sites go below freezing, and retreat to deeper locations where the temperature is more stable and they can resume their interrupted winter sleep.

There is some evidence that hibernating Little Brown Bats exposed to subfreezing temperatures allow their body temperatures to go below freezing. The evidence is controversial, however, and more research is needed to determine just how often this 'supercooling' approach is used. If it were common, many more sites would become acceptable for hibernation. It could also explain the apparent disappearance of so many bats between summer and winter roosts.

One advantage of hibernating in an exposed site is the chance to arouse and feed during warm periods in winter. Although there are relatively few such warm spells in north temperate regions, temperatures may often climb to well above 10° C during winter in more southern parts of the temperate zone. Above

Changes in temperature in an abandoned mine in Ontario have resulted in condensation of water droplets on the fur of this hibernating Little Brown Bat (8 grams). At the time this photograph was taken, the temperature in the mine was 3°C, the same as the temperature of the bat. Outside it was –20°C.

10°C, flying insects become quite common, and an insectivorous bat active under these conditions would have a chance to replenish its energy supply before returning to its hibernation site. When food is available, Gould's Long-eared Bats in Australia arouse from their torpor each night of winter to feed; when food is scarce they go into a more prolonged sleep and make no effort to feed.

Red Bats, species assumed to hibernate in hollow trees in more southern parts of North America, do not arouse when the temperature dips below freezing. Instead, they raise their metabolic rates just enough to keep from freezing and remain torpid until the outside temperature goes above 20°C. This thermostat setting means that Red Bats cannot hibernate in regions with prolonged sub-freezing temperatures where they would burn themselves out. Nor can they use well-insulated sites in caves or mines as hibernacula since temperatures there do not rise to 20°C, denying the bats their cue for arousal.

Bats hibernating in underground sites are more insulated from day-to-day changes in temperature and are less able to exploit favourable breaks in winter's grip. For the same reason, of course, underground sites offer stable temperatures, often around the annual mean temperature of the region, and fewer disturbances to hibernation. Some locations within underground sites are more insulated from outside conditions than others, and their selection as hibernation sites reflects the particular requirements of different species. In southern Europe, for example, Blasius' Horseshoe Bat does not hibernate in any location where the temperature falls below 14°C in midwinter. In Ontario, Small-footed and Big Brown Bats select sites in mines and caves where air movement produces fluctuations in temperature and lower humidity. These species have a high tolerance for temperature change, remaining in hibernation between 10°C and –9°C, and arousing above and below this range. Some species are much fussier. Little Brown Bats usually pick hibernation sites with temperatures between 1°C and 5°C in Ontario, but they will tolerate temperatures as low as –4°C. Temperature changes which cause Little Brown Bats to arouse from torpor vary according to their geographic distribution; in southern parts of their range, such as Georgia, these bats may continue to hibernate at 10°C, a temperature that would induce arousal in an Ontario population.

Humidity is also important to hibernating bats. The levels of relative humidity in bat hibernacula are often over 90 per cent, and efforts to maintain captive bats in hibernation usually fail unless the air is almost saturated and the temperature conditions are just right. High levels of humidity appear to prevent dehydration, and clustering behaviour in hibernation may also be related to minimizing water loss.

By the end of their hibernation period, bats have lost up to 25 per cent of the weight they had the previous autumn, and their fat reserves are severely depleted. The factors immediately responsible for emergence from hibernation are not clear, but it is probably a combination of depletion of body fat and natural arousal that puts the torpid bat back into action.

Arousal consumes more energy than any other facet of hibernation. Hibernating ground squirrels arouse from hibernation about every 10 days, using these brief periods of activity to eat and empty their bladders. These actions are necessary because even during hibernation ground squirrels burn some energy and must get rid of the resulting wastes and replenish their energy supply. Studies show that the animals eat just enough food to compensate for the cost of arousal.

How, then, do bats survive the winter on their meagre store of body fat? Studies of Little Brown Bats have provided some answers. These animals can go as long as 90 days without arousing from hibernation, avoiding the most expensive part of the process – waking up. We still do not understand how bats can remain torpid for so long, or how they avoid poisoning themselves with their metabolic wastes, or what factors eventually wake them for spring. There is some evidence they may awake under the nervous strain of a full bladder; following arousal, bats usually urinate, groom themselves, and return to torpor, sometimes after a short flight.

At one time, some biologists believed bats were poor regulators of their body temperatures because of their propensity to enter torpor in the summer. However, this interpretation failed to take into account the fact that relatively few species of bats become torpid, and it ignored the fact that there is a fundamental difference between the torpor of a bat in hibernation and that of a bat in summer. This difference is simple: the winter bat can spontaneously arouse from torpor of hibernation, while the summer bat requires external stimulus. Evidence of spontaneous arousal is provided by the annual departure of bats from hibernacula where neither light nor temperature conditions offer cues for arousal. At one site in an abandoned mine in Ontario, a thermometer that records maximum and minimum temperatures showed less than a 1C° change over five years. Furthermore, those supporting the view that bats were sloppy temperature regulators failed to recognize that summer torpor allows bats to avoid situations in which they would have to burn energy to remain active. From an ecological point of view, bats' strategy of body temperature regulation is excellent for a small animal, and one shared to some degree by hummingbirds and some moths and bees.

The beginning of the hibernation period in Little Brown Bats closely follows the start of the mating season, suggesting that the bats' systems are controlled to some degree by changing day length. This type of environmental control is common among living organisms, from plants to animals. Longer nights and shorter days appear to activate the reproductive system and initiate the changes in physiology and behaviour appropriate for hibernation.

The sight of a television reporter or biologist handling a hibernating ground squirrel cannot fail to convince the viewer that the animal is oblivious to its surroundings. Hibernating bats are very aware of their surroundings, however, and they may be aroused by sound and touch as well as by changes in temperature or humidity. Indeed, a disturbed bat requires only 20 or 30 minutes to become fully active, which means it has warmed up at a rate of about 1C° per minute. This sensitivity allows bats to respond quickly to changes in their hibernation sites, and may permit survival in the face of drastic drops in temperature. It also underscores the importance of stable, well-protected roosts. As cave exploration becomes a more popular pastime, disturbances of hibernating bats increase accordingly. Even considerate cave enthusiasts who do not handle or even closely approach hibernating bats are still likely to waken them and thus destroy the pattern of careful arousal rationing that is the key to their survival.

Besides questions about torpor and arousal, several other facets of bat hibernation remain poorly understood. How do bats locate suitable sites, and, having found a cave or mine, how do they determine if the site will be appropriate in midwinter? Many caves and mines that look as though they should harbour hibernating bats do not. Equally puzzling is the speed at which bats find and occupy new sites. In one mine in Quebec, bats were seen only occasionally underground during the period from when it opened in 1895 until the late 1950s when mining operations broke through to the surface at one place. This opening provided, for the first time, large direct access to the mine, and by 1969 it was home for several thousand hibernating bats of five species. Five years later, however, the bats were gone, for after the mining operations ceased, the workings flooded. Presumably the bats found an alternative site, or returned to the places they had used before the mine became accessible.

The role of clustering behaviour in hibernating bats is another tantalizing question still unanswered. During hibernation, some bats gather in dense groups that may include thousands of individuals, while others hibernate alone or in small, loosely packed clusters. Little Brown Bats practise a full range of behaviour, sometimes hibernating alone, other times in small groups, and often in huge tightly packed clusters.

As we have seen, the functions of clustering in summer day roosts seem obvious: it serves to maintain high temperatures of the bats forming the cluster, and in some cases acts to raise the temperature of the roost. In some situations, clustering by hibernating bats may have a thermal function and perhaps serve to keep bats from freezing. In many other cases, however, the temperatures of the cave rock, the surrounding air, and the bats are identical, effectively ruling out temperature manipulation as a reason for clustering. It is possible that clustering regulates the rate at which bats lose water, although it is difficult to explain water loss in sites where the relative humidity approaches 100 per cent. It is interesting to note that by the end of the winter Little Brown Bats hibernating in clusters weigh significantly more than those hanging alone in the same cave; perhaps clustering makes bats less vulnerable to water loss simply because it reduces the amount of body exposed to the surrounding air. Social factors may also influence the clustering habits of some bats during hibernation. Male Little Brown Bats tend to hibernate in clusters, while most females spend the winter alone or in small groups. However, the reasons for these differences in behaviour are not known.

Bats use two kinds of fat, both common to many mammals, to help them through hibernation. White fat is well supplied with blood vessels and serves as an energy reserve and an insulator. This is the fat that plagues dieters and is used by whales and dolphins to insulate themselves from cold water. Brown fat, on the other hand, serves not only as an energy store but also as a space heater. Common among young and hibernating mammals, brown fat has a high capacity for producing heat. During arousal from hibernation, the brown fat gland heats up the blood passing through it to speed the waking up process. These glands vary in size according to seasonal needs; in Big Brown Bats the brown fat gland is large in midwinter and small in summer.

Hibernating bats also have specialized enzymes that permit normal bodily functions at cold temperatures when normal body enzymes would be incapacitated.

Because most biologists with access to the resources needed to conduct physiological research into animals' overwintering strategies live in temperate areas, we know most about overwintering in temperate bats. It is important to remember, however, that in many tropical countries, winter is a prolonged dry period, perhaps accompanied by very hot temperatures; a time of shortage of insects. Under these conditions some bats endure food shortages by living off fat they have stored on their bodies, usually in the membrane between their hind legs (the interfemoral membrane). The period of endurance and general inactivity, referred to as 'aestivation,' is a strategy of other animals besides bats. Several species of tropical bats have been reported to spend up to two months in aestivation, resting in sites such as caves or tombs that are well sheltered from the extremes of heat. Unlike hibernating bats, the aestivating ones are active and alert, not requiring 20 or 30 minutes to respond to a disturbance.

The ability of some bats to raise and lower their body temperatures is one of the keys to their success and explains the success of plain-nosed and horseshoe

Hibernating Indiana Bats (8 grams) form densely packed clusters in caves throughout their range in North America. The function of these clusters is not clear. Banded individuals permit biologists to study the longevity and movements of bats.

bats in temperate areas. Their variable thermostat permits efficient use of energy over a wide range of climatic conditions, and helps to explain patterns of activity that have been reported from different times of the year. For example, a visit in May to a cave in Canada used as a hibernation site by bats would probably reveal only an occasional stray Little Brown Bat, probably a male. At the same site from May until the beginning of August, bats would be rare during the day, but on an August evening hundreds of bats would fly in and around the location. By September, visits during the day would reveal hibernating bats in increasing numbers, and from then until May they would be consistently present.

Whether they have a variable thermostat or not, appreciating some of the energy constraints imposed upon bats makes it easier to understand their selection of roosts, their patterns of activity during the night, and seasonal changes in their distribution and behaviour. This should become increasingly obvious as we forge ahead and ponder the use bats make of roosts.

Roosts

Although bats usually pick caves, rock crevices, or different parts of trees for their roosts, they will also use mines and buildings, and some occupy what seem to us bizarre roost sites. I can recall spending a morning crawling on hands and knees elbow-deep in mud through a Warthog hole in Zimbabwe in search of Hildebrandt's Horseshoe Bats. Warthogs spend their nights in these cave-like tunnels, which are also used by other animals, including hyenas, porcupines, a selection of snakes and lizards, and, of course, bats. Armed with a hand net and headlight, I slithered into the hole behind Naboth Tsindi, a game scout with Zimbabwe's Department of National Parks and Wildlife. As we proceeded along the passage, Naboth turned to advise me that if a Warthog appeared, I should do my best to get out of the way. Fortunately none did, but the multitudes of fleas and ticks awaiting their return appreciated our presence. The horseshoe bats proved to be more elusive, and several hours of careful stalking and frantic netting produced only one capture.

The story serves as an example of the ingenious opportunism of bats looking for roosts, and an illustration of the ways in which different animals will share suitable sites.

The design of bats, especially the arrangement of their flight muscles, which permits them to squeeze into tight locations, suggests that roosts have always been crucial to their survival. They provide sanctuary from predators and refuge from excess heat or cold. Roosts are usually divided into three broad categories, according to their role in the lives of bats: day roosts, night roosts, and hibernation sites. Day and night roosts serve as temporary refuges for active bats, hibernacula are occupied during long periods of cold temperatures.

DAY ROOSTS

Bats can be separated into three general types according to whether they choose hollows, crevices, or foliage as day roosts. Hollows may be found in caves, mines, trees, or open spaces in buildings. Crevices may be found in rocks or buildings or under the loose bark of trees. Foliage provides almost unlimited roosting opportunities, and some bats hang up under leaves, from branches, or against the trunks of trees. However, this generalized classification conceals a number of interesting variations; the bats' choices of day roosts are almost as varied as the animals themselves. In fact, the only common denominators for all day roosts are shelter and appropriate temperature, and even this statement could be challenged on the basis of shelter.

Bats using hollows as day roosts are usually considered pests if they select a building as their refuge. However, bats roosting in caves or hollow trees are not nearly so conspicuous. Indeed, their hollows are often small and remote and finding them can present the biologist with a challenging problem.

As the story of the Warthog hole illustrates, roosting hollows are often not the exclusive retreat of bats. In Africa, Egyptian and Wood's Slit-faced Bats sometimes roost in caves that also house huge colonies of bees, an association that provides the bats some protection from disturbances. In Malaysia, several species of bats share caves with huge colonies of small swifts known as swiftlets, and the entrances to these caves are often the sites of frantic traffic jams at dawn

These sketches show some of the places where one commonly finds roosting bats in the vicinity of buildings, rocks, and trees and other foliage. The bats, shown as small dark blobs, may be conspicuous or well hidden, and clearly exploit a wide range of roosts. Sketch by Connie L. Gaudet.

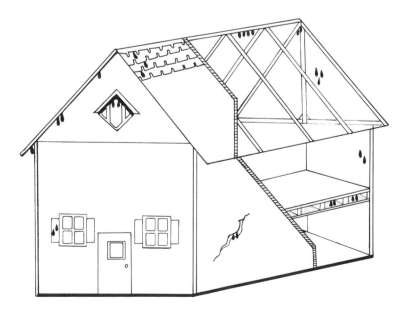

and dusk as the swiftlets leave and the bats return and *vice versa*. On the other side of the world in Kansas, a colony of Big Brown Bats and a nesting pair of Barn Owls shared the attic of an old building. Owls often take advantage of these situations, common in many parts of the world, by using the bats as a handy food supply. In Africa, the huge hollow trunks of baobab trees frequently harbour bats during the day and birds at night. In one boabab tree about 25 metres in diameter, I found a huge colony of Hildebrandt's Horseshoe Bats, some Common Bent-winged Bats, a few swifts and a pair of Barn Owls.

Bats occasionally roost in the nests of birds, usually after they have been abandoned. Woolly Bats, a little-known group of plain-nosed bats which occur from Africa to Australia, often settle into the nests of weaver-birds. In Arizona, Yuma Myotis, a species similar to Little Brown Bats, move into the mud nests of Cliff Swallows after the birds have moved out.

Much less is known about the room-mates of bats that roost in small hollows, crevices, and foliage during the day, partly because the roosts are difficult to locate and enter. In Malaysia, however, two species of Club-footed Bats roost in

the hollow stems of bamboo, getting into these sanctuaries through small holes made by boring beetles. These bats have very flattened heads which allow them to squeeze into the holes made by the beetles. Unfortunately, there appears to be no information about the evolutionary history of the relationship between the Club-footed Bats and the boring beetles.

Some bats that roost in crevices also have spectacularly flattened skulls, particularly several species of free-tailed bats that squeeze under rocks in the savannah woodlands of Africa and South America. Two of these species, the South American Flat-headed Bat and Peters' Flat-headed Bat from Africa, are further adapted for tight places by wart-like growths on their forearms that apparently protect their delicate skin from abrasion against the rocks under which they roost. Interestingly, the third species, Roberts' Flat-headed Bat from Africa, first came to the attention of biologists after it was caught by people overturning rocks in search of scorpions. Even bats not noticeably specialized for roosting in crevices can squeeze into small openings, a fact well known to homeowners trying to bat-proof their residences.

This South American Flat-headed Bat (6 grams) roosts under rocks in the savannahs of tropical South America. Its flattened skull reduces its profile, permitting it to squeeze into small openings. Sketch by Paul Geraghty.

Bats that hang up in the leaves of trees take one of two approaches to life in their roosts. Most species are cryptic, blending in with the background of their roosts, and these are rarely encountered. However, some flying foxes form large camps that are easily detected, since the bats make no effort to conceal themselves, either in their choice of roost site or in their behaviour; they are both noisy and active. But whether their habits are cryptic or conspicuous, bats roosting in foliage are active and alert, quick to take flight at the approach of danger.

Foliage-roosting bats include representatives from many families. In North America, Red, Hoary, and Yellow Bats roost in foliage, often hanging under leaves. In Central and South America several species of sheath-tailed bats roost on the trunks of trees, and many leaf-nosed bats roost among the leaves. In Africa, Asia, and Australia, most flying foxes and their relatives hang in trees, along with many sheath-tailed and plain-nosed species. Although some African

Wedged between the brick of a chimney and a roof board, these Little Brown Bats demonstrate their ability to squeeze into crevices, an attribute well known to people trying to bat-proof their homes. This crevice is about 10 millimetres wide.

Dog-faced Bats roost in caves, most of the flying fox relatives in Africa roost alone or in small inconspicuous groups. The exception is the Straw-coloured Fruit Bat, which often forms large, conspicuous colonies numbering thousands of individuals.

The 'real' flying foxes, bats in the genus *Pteropus*, often congregate in large, noisy camps. In India, where bats are protected to some degree by religious custom, the camps of flying foxes may be found amid the heavy traffic and bustling activity of city centres. In Australia, where people have a different view of bats, flying foxes are often less brazen, and frequently live in more remote areas.

I once visited a camp of Grey-headed Flying Foxes on the edge of a golf course in Brisbane. The trees were festooned with these bats, which seemed to be perpetually bickering among themselves and changing position from one tree to another. Any effort to approach them closely caused an immediate exodus from the vicinity, a retreat to more secluded trees. In comparison, the behaviour of a camp of Black and Grey-headed Flying Foxes near Chillagoe in north Queensland suggests that when these bats less commonly encounter people they may be less nervous. The camp near Chillagoe was located on the edge of a limestone tower. Getting there required an arduous climb through a belt of thorn thicket, and then up and over razor-sharp rocks that made up the tower's sheer walls. The visit was worth the effort. I was really impressed by the sight of so many bats clinging to the rock surface, and making clumsy landings on the sharp rock, apparently without damaging their wings. It was amusing to see bats that usually hang free from branches living with their backs to the rock walls. I have no idea how much, if any, persecution had driven these flying foxes to such an uncomfortable roost. But in this isolated and difficult-to-reach location, the bats were less wary of observers, and provided a first-class demonstration of flexibility in roosting behaviour.

One foliage-roosting species, the Wrinkle-faced Bat from the Neotropics, has developed an interesting adaptation that continues to puzzle biologists. The wing membrane between its second and third fingers is transparent, and the bat positions these windows over its eyes when roosting with its wings around its body. The windows allow the bat to survey its surroundings for approaching danger, and may help to hide it from predators by concealing its eyes.

Of the known specializations for roosting by bats, those of the tent-making and disc-winged species stand out from the rest. The tent-makers, which include several species of New World leaf-nosed bats, and at least one flying fox relative, the Sphinx's Short-faced Fruit Bat, are the only bats whose roosting habits approach the nest-building abilities of birds. To construct their tents, Honduran White Bats from Central America use their teeth to cut the veins running out from the mid-rib of long, broad leaves, such as banana leaves. This causes the sides of the leaves to fold down in the shape of a tent, protecting the bats from sun and rain (either of which would drastically alter the bats' energy balance). Honduran White Bats often roost in small groups of four to six individuals, and, perhaps as a means of avoiding predators, the group rarely spends two consecutive days in the same tent, preferring to rotate among several of these shelters. The leaves are usually about two metres above the ground, and hang well out from the plant – out of reach of most four-footed predators. Furthermore, the stems of the *Heliconia* leaves they often select for their tents are not strong enough to support most predators, and even animals small enough to try to clamber up to the bats cause enough vibrations to alert the bats to the approaching peril.

Tent-making bats are not the only species that move the locations of their roosts from day to day. On a visit to Zimbabwe in early 1982, I was able to follow the roosting habits of Lesser Yellow House Bats by attaching small radio transmitters to their backs. Over a one-week period, the bats rarely spent two consecutive days in the same roost, usually the upper reaches of hollows in mopane trees. Other species, including Big Fruit-eating Bats and White-lined

Tailless Bats from Panama, also routinely change their foliage roosting sites.
But many other species of bats occupy the same roosts day after day, year after
year. The reasons for the two different roost-occupation patterns in bats are not
clear, but switching unpredictably from one location to another may serve to
foil predators. There is no evidence that moving from site to site is related to the
type of roost occupied, for bats roosting in foliage and hollows may be home-
stickers or movers.

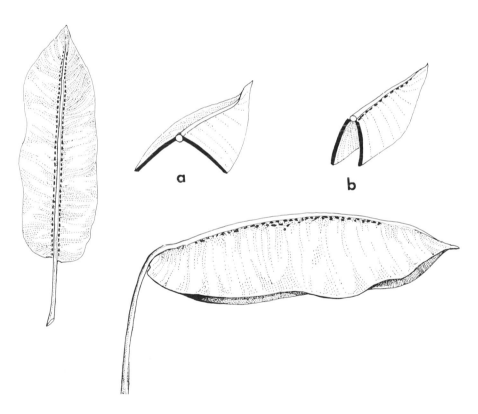

This *Heliconia* leaf has been made into a tent by a Honduran White Bat. By biting
(dashed lines) along the leaf on either side of the centre vein, the bat causes the leaf to
fold down on itself (compare before and after pictures, a and b respectively),
providing protection from sun and rain. Sketch by Connie L. Gaudet.

Disc-winged bats have adhesive discs on their wrists and ankles that permit them to move around freely on the smooth, waxy surfaces of leaves. Three species comprise this group, which includes two families. One species, the Old World Sucker-footed Bat, occurs only in Madagascar. Spix's Disc-winged Bat and its close relative the Honduran Disc-winged Bat occur in Central and South America; they are not considered to be closely related to the Madagascar form. Most of our knowledge about disc-winged bats is based on a study by James Findley and Don Wilson, two American biologists who were working in Costa Rica. They studied Spix's Disc-winged Bat and found that the animals roosted in unfurled *Heliconia* leaves, attaching themselves to the slippery leaf surface with their discs. The unfurled leaves were tube-shaped, and the bats only roosted in them when the end opening of the tube was 50 to 100 millimetres in diameter. Since unfurling leaves only have openings of this size for one day, Spix's Disc-winged Bats must find a new leaf every day. Groups of six bats usually occupy one unfurled leaf, and as a rule, most groups include one adult male. In the laboratory these bats often licked their discs, and the degree to which these suckers are specialized to cling to smooth surfaces could be seen in the objects they were able to cling to. Glass and the fingers of observers were easily mastered, but wood and window screen were not. When several bats were placed into a tin can, the popping sounds made by their discs as they were disengaged from the metal were reminiscent of many bottles of wine being opened at the same time.

Although specializations such as adhesive discs, flattened skulls, and wart-like growths provide excellent illustrations of the ways bats exploit secure day roosts, the speed at whch they find and occupy new roosts and their flexibility in roost selection are equally astounding. In many cities, for example, the levels of bat activity are higher than in the surrounding countryside. Presumably this reflects the ability of bats to take advantage of the wider variety of roosts available in buildings.

NIGHT ROOSTS

Our knowledge of bats' night roosts remains largely incomplete; indeed for many species no information is available on this topic. We do know, however, that many insectivorous and frugivorous bats use night roosts for a variety of activities, including eating and grooming. Biologists usually obtain clues to the presence of night roosts from bits of discarded fruit or insects and piles of droppings in locations where bats are not seen during the day.

In eastern North America, Little Brown Bats often gather in groups of more than 50 in the holes in barn beams during the night. In these confined sites, the collective heat of the bats warms the roost, ensuring rapid digestion of food and

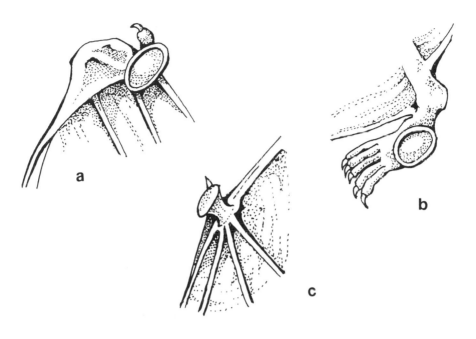

Adhesive discs permit some bats to move around on very smooth slippery surfaces. The discs are developed on the wrists (a and c) and on the ankles (b), and are best known from New World (a, b) and Old World Disc-winged Bats. The discs on the wrists of the Old World Disc-winged Bat are stalked (c), while those of the New World Disc-winged species are not. Sketch by Connie L. Gaudet.

release of the energy it contains. Not all species make their night roosts in confined spaces. In Lake Eacham National Park in Australia, an Eastern Horseshoe Bat made its night roost over the toilet in the men's washroom, usually hanging from the ceiling by one foot. In this instance, bat droppings on the seat provided concrete evidence of a night roosting site.

Night roosts may harbour solitary bats, groups that live together in day roosts, or a female and her young. But studies of Little Brown Bats in southern Ontario revealed no obvious social structure to their night roost populations. It is possible that night roosting allows one bat to learn from others the locations of good feeding sites, but this is a suggestion much easier to present than to prove experimentally.

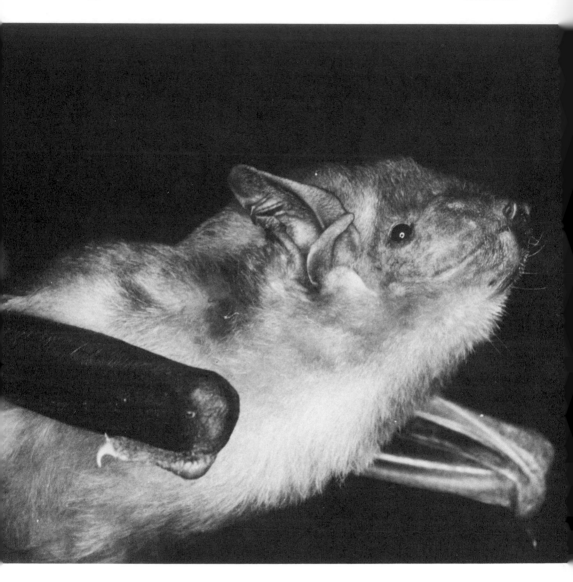

This Lesser Yellow House Bat (20 grams) from Zimbabwe roosts in hollow trees, frequently moving from one tree to another, apparently to confound predators.

Although there is little hard evidence to show how bats divide their time between feeding and roosting, they appear to spend the larger part of their time roosting. Certainly, some insectivorous species are active throughout the night, although they may intersperse periods of feeding with spells of roosting. But our studies of Lesser Yellow House Bats showed that during every 24-hour period, these bats hunted for one hour, a period of continuous flight. However, I suspect that had we been studying lactating females, we would have found that more time was spent feeding.

The retreat of bats from fruit trees to adjacent sites where they consume their food could be a defence against predators that might lurk at the feeding locations. In one night, a bat may make numerous trips from fruit trees to night roost, and there is often a large accumulation of seeds under the night roosts; this may be important in the role of dispersal of the food plants. In fact, the seeds of some plants show higher rates of germination after they have passed through a bat's digestive tract.

HIBERNATION SITES

As we have discussed, bats enduring prolonged periods of unfavourable weather usually select roosts on the basis of temperature and humidity. For bats avoiding continuous subfreezing temperatures, suitable hibernacula may be caves, abandoned mines, storm sewers, and basements with cave-like conditions. In somewhat warmer climates, bats may hibernate in hollow trees, under pieces of loose bark, or in cracks and crevices in rocks. The range of hibernation sites available declines with the increasing severity of winter, particularly with prolonged subfreezing temperatures. In the Netherlands, for example, where winter does not bring long periods of subfreezing temperatures, Noctules hibernate in hollow trees. On the other hand, bats in most of Canada face long, bitter winters and are not known to hibernate in trees. Instead, Canadian bats select underground sites where humidity is high and temperatures usually remain above freezing.

The adaptations of bats provide strong evidence of the vital importance of roosts to their survival. Indeed, the speed at which they find new roosts, often to the despair of homeowners, suggests that the availability of roosts can affect the size of bat populations. This is clearly true for some populations of Spix's Disc-winged Bats, which, because they occupy their day roosts for only one day, face limited numbers of sites. Furthermore, bats are very sensitive to disturbance and will often abandon a compromised site, a response that has important implications for their survival.

Activity

Seasonal changes in the levels of bat activity are directly related to temperature conditions. One evening in May 1976, I sat outside the Rockefeller University Field Research Center near Millbrook, New York, waiting for the nightly exodus of the resident colony of about 30 Big Brown Bats. The wind was tart, and the mercury hovering around the 3°C mark, about 8C° cooler than the temperature inside the bats' roost. The weather did not bode well for the bats, and I looked forward to observing their response to the wind and cold. A bat soon emerged from the roost, followed in quick succession by about a dozen others. They each made one circuit of the adjacent parking area, and quickly returned to their roost for the remainder of the night. Their actions stood in sharp contrast to warmer nights when the entire colony usually emerged at the same time of evening, and stayed out for the entire night. Clearly, the drastic decline in the bats' activity on this cooler evening was related to a similar decline in the numbers of flying insects at temperatures below 10°C.

For at least two reasons, levels of bat activity also decline when it is raining, although lactating females often leave roosts in search of food even during thunderstorms. Echolocation may be less effective in the rain because the atmospheric absorption of high-frequency sound is increased by high humidity, and raindrops could produce extra echoes which might confuse the bats. A more reasonable explanation involves the bats' temperature regulation. Wet fur and wings would increase the amount of evaporative cooling to which the bat was subjected, upsetting the animal's temperature regulation and energy balance during flight.

Disturbances, whether they be natural or artificial, also affect levels of bat activity. We have seen how disturbances to hibernating bats will lead to their arousal from torpor, and such disturbances can result in movements of bats between hibernation sites in midwinter. Banded Little Brown Bats have moved more than 120 kilometres between caves in midwinter, perhaps because of the agitation caused by banding them. Small-footed and Big Brown Bats regularly leave their midwinter hibernation sites when outside temperatures climb above the freezing point. A cave that was full of these bats when the outside temperature was –20° C may be empty the next day if the mercury has climbed to 5° C. We do not know where they go or why, but their absence from hibernation sites speaks for itself.

In spring, summer, and early autumn, the activity of bats away from their day roosts is sometimes sporadic, especially during cold weather. But under appropriate temperature conditions (above 10° C) bats usually leave their day roosts when it gets dark. Several studies suggest that changes in light levels govern the nightly departure of bats from their roosts. There are many records of bats flying at noon during solar eclipses, for example, and changes in their times of exodus from day roosts over a season coincide with changes in the timing of dusk. Other studies have revealed that day-to-day changes in the pattern of departures often are related to differences in cloud cover. A demonstration of the role of light intensity in the timing of bat departures from day roosts was provided a few years ago when I was demonstrating that the installation of lights in an attic would stimulate the bats to move out. Before abandoning the lighted sites, the bats began emerging from them significantly earlier than they did from neighbouring colonies without lights. To bats in a lighted colony, it appeared darker outside earlier than it would to bats in a dark colony: changes in relative levels of light seem to give the animals cues for the timing of their departure.

Moonlight also depresses the activity of bats. Two explanations for this lunar phobia are tempting, but neither is proven. Moonlight could make it easier for predators such as owls and hawks, which see better in brighter light, to find and attack flying bats. However, with the exception of the tropical Bat Hawk, we know of no predator that specializes to any extent on bats. Moreover, even bats from areas where no predators are evident are less active in bright moonlight. Bats' avoidance of moonlight might also be related to a decline in the abundance of insects. Anyone who has used lights to attract insects will have noted fewer insects drawn to their lights when the moon is bright. However, lack of insects at artificial lights does not prove that there are fewer insects about to be caught by hunting bats, nor does it explain the lunar phobia of fruit-eating bats.

During a series of studies in Zimbabwe, I found that in bright moonlight insectivorous bats hunted for insects under the tree canopy, spending almost no time above the trees or over open grasslands. On dark nights, the bats hunted mainly in open areas, and above the tree canopy. A pair of Bat Hawks lived near the study site and their diets reflected the general abundance of different kinds of bats in the area. It is therefore possible that the bats behaved in a less conspicuous fashion in bright moonlight to avoid the Bat Hawks, although we did not see the birds in our immediate study area. As a footnote, it is worth noting that the bats caught the same kinds of insects on bright and dark nights.

Levels of bat activity away from roosts change throughout the night. Often, definite peaks of activity are obvious, perhaps reflecting bouts of feeding, and the captures of bats in mist nets or traps suggest that these bursts of activity occur most often at dusk and at dawn. By monitoring echolocation calls, however, it becomes clear that some bats are active throughout the night. Only studies of individual bats, perhaps those carrying small radio transmitters, will clarify how they budget their time between feeding, night roosting, and other activities. It seems obvious that different bats, adults, subadults, males, or females, may have quite different patterns of activity, reflecting their different physiological states and demands.

Migration and Navigation

Bats in many parts of the world avoid unfavourable conditions by migrating to less severe locations. Anyone frequenting the market in Abidjan, the capital of the Ivory Coast, cannot fail to notice the thousands of Straw-coloured Fruit Bats which hang out there from November to mid-February. Although a few of these bats are present for the rest of the year, most of them show a very seasonal pattern of residence. Biologists who have spent long periods in the tropics, especially in areas where flying foxes occur, have remarked that some bats seem to move seasonally from place to place in a predictable pattern. So far, though, no one has tracked individuals from these migrating groups to determine the magnitude of their peregrinations.

Probably the most spectacular of the documented bat migrations is the annual movement of millions of Mexican Free-tailed Bats between the southwestern United States and parts of Mexico. Large-scale banding operations at caves in the U.S. where the bats spend their summers and subsequent recovery efforts in their winter roosts in Mexico have revealed migrations of almost 1,300 kilometres, one way. The large populations of insects in Mexico provide the bats with a continuous food supply in winter. If you remember that each of these 12-gram bats eats between 20 and 50 per cent of its body weight every night, it is easy to appreciate the size of the insect supply required to feed millions of them.

Many other species also travel long distances between their summer and winter homes. In Australia, some Common Bent-winged Bats are known to migrate more than 1,000 kilometres between summer and winter roosts, and in Europe, the Noctules may move similar distances. In North America, Little Brown Bats regularly migrate more than 200 kilometres from winter to summer roosts, although the record exceeds 800 kilometres. The bat in question was banded in December 1966 in an old mine near Schreiber on the north shore of Lake Superior, and captured the following August at a more southern location near Renfrew, Ontario. A month after its capture at Renfrew, it had returned to the Schreiber location and was deep in hibernation. Like this bat, many species in the northern hemisphere will travel north to find a hibernation site; in other words, bat migration is not always a move towards the equator as cold weather approaches.

Donald Thomas took this photograph of a cloud of Straw-coloured Fruit Bats as they arrived at the Lamto Ecology Station in the Ivory Coast in February 1979. They were part of a migrating group of about 100,000 bats that stayed in the area for 30 to 40 days while the Iroko tree (a tropical hardwood) was in fruit. Some of the females in this group were carrying their babies with them.

Mexican Free-tailed Bats (12 grams) occur
widely in North, Central, and South
America, and are noted for their extensive
migrations between nursery colonies (often
numbering millions of individuals) and their
winter ranges. Sketch by Paul Geraghty.

Since the 1940s, biologists have been puzzled by a mystery involving disappearing Little Brown Bats. Detailed studies in New England, Indiana, Kentucky, and Ontario all revealed that very few females and young banded in nursery colonies in summer were found in hibernation sites in winter. Most of the bats encountered in hibernation sites were males. Between the midsummer break-up of the nurseries and the initiation of hibernation in autumn, females and their young appeared at caves and mines which would serve as hibernation sites for other bats later on, but they usually did not use these locations for hibernation. There are many possible explanations. For example, the bats might be hibernating in the burrows of rodents such as woodchucks, or they might have moved into parts of caves not accessible to people. On the other hand, if Little Brown Bats can tolerate supercooling as some experiments have suggested (see page 73), females may be using hollow trees or other exposed sites to pass the winter. The answer to the puzzle will only be known when we are able to follow these bats to their hibernacula. In the meantime, one becomes used to studying female and young Little Brown Bats in summer and adult males in winter.

Migrating bats do not necessarily make long journeys between their summer and winter roosts. In North America, for example, most records for Big Brown Bats indicate movements of less than 40 kilometres, and the 230-kilometre long-distance record for this species appears to be exceptional. These shorter movements are no doubt related to the Big Brown Bats' greater tolerance for cooler, drier hibernacula relative to other species, providing them with a wider range of sites from which to choose.

Between the time nursery colonies disperse in midsummer and the onset of weather unpleasant enough to induce hibernation, bats of temperate areas become vagrants. This behaviour coincides with the start of the mating season and migration, and frequently results in bats appearing in buildings or on ships at sea; vagrant bats often turn up in the most unexpected places! There are, for example, occasional records of Hoary Bats from Iceland, one from the Orkney Islands, and frequent records of them and Red Bats from Bermuda.

Red Bats (12 grams) commonly roost in the foliage of trees and bushes. This male was very cryptic in his roost in a crabapple tree, a stopover on his fall southward migration.

NAVIGATION

We know that migrating birds rely on celestial, magnetic, and other cues to navigate over long distances, but much less is understood about the orientation strategies of bats. To learn more about how animals find their way, researchers often monitor their movements by using small radio transmitters.

The most detailed study of this kind involving bats was conducted in Trinidad with a colony of Spear-nosed Bats, which were taken from their roosts, outfitted with transmitters, and divided into three groups. One group was untreated except for the radios, the second group was blindfolded and radio-tagged, while the third group wore goggles (blindfolds with the eyes unblocked) and their radios. All the bats were then released at specific distances from 'home' (their cave roost), and their subsequent movements tracked by following the signals from their radio transmitters. Untreated and goggled animals freed 9.5 to 16 kilometres from home flew almost straight back to their cave, but the blindfolded bats dispersed randomly, and took erratic flight paths before ending up at home. At 24 to 32 kilometres from home, the untreated and goggled bats took more scattered routes as they headed home, but at 56 kilometres strayed as much as the blindfolded bats at 9.5 kilometres.

The results suggest that bats have what might be called a 'familiar area,' the range they regularly traverse and know intimately. For Spear-nosed Bats, the familiar area extended less than 56 kilometres from home. The study also shows that vision plays an important role in bat navigation, although the blindfolded bats did eventually find their way home. When we consider that echolocation is a relatively short-range system of orientation (see page 33), it is not surprising that vision would be more important for general orientation.

'Homing studies,' those involving displacement of an animal from its home to some distant location, are a common means of studying navigation. However, most homing studies of bats are hampered by two limitations which have led some biologists to question the whole basis of this approach. First, the bats often move to different roosts after being disturbed. In studies where the researchers score homing by bats returning to the location where they had been captured, the movement of a bat to another roost (an alternate home) would appear to count as an unsuccessful attempt to find the first home. The Trinidad study of Spear-nosed Bats avoided this problem by equipping their bats with radio transmitters, an approach not feasible for many smaller species of bats.

A second problem with studies of homing by bats deprived of their vision, in some cases by blindfolds, in other cases by blinding, involves vulnerability to predators. Low success rates of homing by blinded or blindfolded bats may reflect their inability to tell day from night. Several observations suggest that day-flying bats are particularly vulnerable to attacks by hawks, so the experimental bat deprived of its vision and not coming home may have not told the experimenter anything about its navigational ability.

So, while it is clear that vision plays some role in bat navigation, we still have much to learn about the orientation skills of bats. Recent work indicates that there are deposits of magnetic material in the brains of bats, implying a possible magnetic sense. As the technology to build smaller radio transmitters improves, more species of bats will be eligible for homing studies, and our potential to monitor their considerable navigational skills will be increased. Although their travels are not nearly so spectacular as those of some birds – the pole-to-pole journeys of Arctic Terns, for example – many bats successfully find their way home over long distances year after year. As usual, the challenge is to the biologist to come up with a suitable way of finding out how they do it!

Reproduction

People often call bats 'flying mice,' and assume that, like mice and other rodents, they reproduce at alarming rates. In fact, bats are not very prolific in this regard; most species have only one or two litters per year, each with only one or two young. Red Bats produce the largest litters known from bats, occasionally having four babies, but more often they have only two or three, and they only have one litter annually. Even the most fertile of bats produce a maximum of four newborn a year.

Some species show interesting variations in their patterns of reproduction. For example, Big Brown Bats from eastern North America usually have twins, while their counterparts in the west and in some islands in the West Indies produce only one young per litter. Many other species, including the Serotines or Pipistrelles of Europe, give birth to twins in most cases.

Some bats are able to breed in their first year, while others do not produce young until they are three years old. Most species, however, attain sexual maturity some time between these extremes. Males and females of some species reach sexual maturity at different ages. Female Little Brown Bats, for instance, may give birth in their first year, but males are not capable of inseminating females until they are about 14 months old.

Most bats have a definite breeding season, the period during which they mate, and as a result their young are born within a relatively short time-span in any given year. In temperate regions, birth occurs in late spring, May or June in the northern hemisphere, November and December in the southern. The births of tropical species usually coincide with the start of the rainy seasons, which begin at various times in different parts of the world. Common Vampire Bats appear to be an exception, producing their young at any time of the year. However, even though they may mate at any time, Common Vampire Bats still have only one or perhaps two young in one year because of their relatively long gestation period – five to seven months.

The length of pregnancy for bats ranges from 40 days for small, insectivorous species, to six months for large flying foxes, and nine months for some leaf-nosed bats. The young are comparatively large at birth, and the weight of a newborn bat often exceeds 25 per cent of the post-delivery weight of its mother. For example, a newborn Little Brown Bat usually weighs between 1.5 and 2 grams, while its mother tips the scales at about 8 grams.

Although delivery is often a difficult process, female bats have developed several specializations to accommodate the large size and wing structure of the fetus at birth. The ligaments that hold the pelvic girdle together are very elastic, and stretch to permit passage of the large baby through the birth canal. Furthermore, young are born rump-first, minimizing the chance of their wings becoming tangled in the vagina.

During delivery, some female bats have been observed to turn upside-down (head-up, in other words) to facilitate birth by enlisting the help of gravity. Birth may take up to 30 minutes, and the newly emerged baby is dropped into the interfemoral membrane between its mother's hind legs. The mother licks her baby as it emerges, and once it is out, the newborn bat immediately climbs up and fastens itself to one of its mother's nipples. The female usually eats the afterbirth, and then begins a thorough licking of the baby. Because there have been relatively few observations of births in the wild, most studies involve captive bats, and variations in behaviour of females giving birth may be expected.

Newborn bats grow rapidly, with various parts of their bodies growing at different rates. At birth, the thumbs and hind feet are almost adult in size, and grow very little. However, the forearm and other bones supporting the wing enlarge quickly, producing a wing area in adults that is 10 times the size it is at birth. Like most insectivorous species, young Little Brown Bats rapidly gain weight until they are weaned. But once they switch to insects from their mothers' milk, their body weights decrease as they use up the fat reserves built up during nursing. As we shall see, this decline in weight has important implications for the survival of young bats.

At birth, bats are equipped with distinctive milk teeth that appear to be useless as tools for grinding solid food, but ideally designed for attaching to their mothers' teats. Among insectivorous species, the shedding of milk teeth and the emergence of permanent teeth coincides with flight and the appearance of insects in the diet.

Little Brown Bats, like many smaller species, first take to the air within 18 to 21 days of birth. For larger species, the period until first flight may be much longer, the Common Vampire Bats, for example, may only begin to fly at age 8 to 10 weeks.

Female bats have frequently been observed carrying their young, presumably on their way out to feed. The small bats cling tenaciously to the nipple with their teeth, and grip their mother's fur with their already large thumbs and hind feet. During field studies in the Ivory Coast, I often discovered female Buettikofer's

Fruit Bats and their babies in my mist nets, and in Zimbabwe I watched Gambian Epauletted Fruit Bats arriving at a fig tree to obtain food while their young hung on for dear life. However, because the hunt for flying prey requires considerably manoeuvrability and energy, most insectivorous species do not carry their young with them while they hunt, though most bats, whether they feed on insects or fruit, will move their young from one roost to another when disturbed. Female Greater White-lined Bats, an insectivorous species from South and Central America, remove their young from day roosts to more secluded sites where they leave them during the feeding period. The babies are later retrieved and returned to the day roost with their mother before dawn.

The mating cycle of many bats is typical of most mammals; females are inseminated during copulation, fertilization follows, and the embryos develop into full-sized fetuses in the uterus. However, there are some interesting variations of this cycle, one of which is responsible for the nine-month gestation period of California Leaf-nosed Bats. This species mates in the fall, and fertilization ensues immediately. However, there is little development of the embryo until the following March, when growth accelerates, followed by birth in mid to late spring. The delay in embryonic development ensures that birth coincides with a season when food is abundant enough to maintain lactating females. It also permits mating to take place when both males and females are in prime physical condition and have access to good supplies of food.

Horseshoe and plain-nosed bats from temperate regions have also developed a mating cycle that begins in autumn but still guarantees birth in the spring when there is a predictable food supply. This strategy also occurs in some tropical plain-nosed bats, ensuring birth of the young at the start of the rains. Again, insemination occurs in the fall, but fertilization is postponed until the following spring, so that the technical period of pregnancy (from fertilization to birth) remains quite short. To achieve this delay, female bats store viable sperm in the uterus or vagina, and ovulate several months later. Little Brown Bats, for example, mate in late August and early September, and ovulation occurs in late March or early April when females leave hibernation. The young are born in mid-June (in Ontario), so the gestation period is 50 to 60 days. Although the cues that stimulate ovulation are not well understood, females taken from hibernation in January, kept in captivity at room temperature, and provided with sufficient food become pregnant and give birth several months earlier than females left under natural conditions. However, pregnant females forced to enter short periods of torpor after they have left hibernation show slower fetal growth, providing yet another adjustment to allow timing of birth to coincide with favourable conditions.

After mating, the vaginas of some noctules and horseshoe bats are blocked with plugs made from either vaginal secretions or seminal fluid. It is not clear

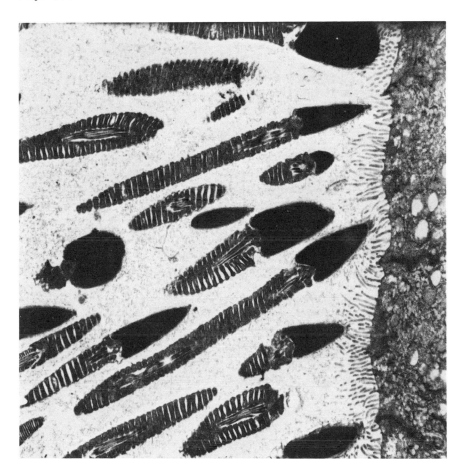

This transmission electron micrograph was provided by Paul Racey. It shows that sperm stored in the uterus of a Pipistrelle are in close contact with the lining of the uterus. Transmission electron microscopy permits tremendous magnification which in turn allows investigation of fine details. This picture is magnified about 6,500 times.

whether the plug is designed to keep the sperm from flowing out of the female tract, or simply as a barrier to unwanted copulations during the period from mating until fertilization.

Although we know that the mating cycles of bats are well suited to their environments, there is still a great deal to be learned about the workings of their reproductive systems. Many studies are required before we will understand just how females store sperm over long periods, or what factors are responsible for initiating ovulation and accelerating fetal growth. Consider the case of female Big Brown Bats. They release up to seven eggs from their ovaries each spring, and all are usually fertilized and the embryos implanted in the uterus. However, many of the developing embryos are resorbed by the female in the period between fertilization and birth, so that the litter size is still one or two, depending upon where the bat lives. How does she accomplish this selective resorbtion?

Perhaps the most intriguing element of bat reproduction involves mating itself. Bats are symbols of fertility in some cultures, probably because male bats boast exceptionally long penises, sometimes decorated with spines. The reason for the long penis is simple and practical: during copulation, the male mounts his mate from the rear, and because many species have broad membranes between their hind legs, the male has a long way to go to meet the female. From the nose to the tip of its tail, a male Little Brown Bat is usually about 14 centimetres long. When erect its penis measures 2 centimetres in length with an impressively large diameter!

Even so, studies have indicated a problem when the penis is inserted into the vagina. Some biologists have suggested that the male inserts the penis before it is erect, and the subsequent enlargement locks it into place. Other observers, noting that copulating pairs are easily separated even when the penis is erect, claim the penis does not act as a locking device and have suggested that it may not be inserted at all. This theory, suggesting lack of intromission, has been advanced for Pallid Bats and Little Brown Bats, but there is no experimental evidence to show that enough sperm could be transferred to the female without intromission.

Many biologists interested in reproduction of mammals find the idea of mating without intromission impossible. More studies are needed to test the theory in question, but it is probably prudent to remember that many ideas previously dismissed as unbelievable have subsequently been shown to be true. Imagine the idea of bats being able to see with their ears!

For more exciting details about the sex lives of bats, see page 124.

Populations

If seeing is believing, then it would not be difficult for anyone to accept that some species of bats achieve the highest population densities of any mammal in the world. Simply stand at the entrance to Carlsbad Caverns in New Mexico (or at one of the other cave colonies) on an evening early in August and you will be an instant believer. As darkness approaches, tens of thousands of Mexican Free-tailed Bats emerge from the depths of the caverns in a chaotic and dizzying display of whirring wings. The seemingly endless cloud of excited animals rushing into the evening air is an unforgettable spectacle. And if you are fortunate enough to have a bat detector at hand, you will also be exposed to an overwhelming cacophony of sound.

At Randolph Air Force Base in Texas, millions of Mexican Free-tailed Bats living in a cave at the end of the runway have forced air traffic controllers to make a few adjustments to prevent collisions with the aircraft using the base. The clouds of bats boiling up from the cave in the evening and returning in the morning show up clearly on the radar screens at the base, and the controllers use this information to schedule the landings and takeoffs of aircraft around the comings and goings of the bats. The emerging bats climb to altitudes of more than 3,000 metres before dispersing to feed, and on their pre-dawn return some reach altitudes of 4,500 metres. The reasons for this behaviour are not clear. While the radar picture tells us the bats climb to these altitudes it does not indicate what the animals are doing. Presumably, some are chasing insects.

Biologists trying to determine the condition of a population of animals benefit from knowing its age structure. We need to know what percentage of the population is in each age group, and how many individuals are capable of reproducing at any given time. The size of the reproducing population allows us to distinguish declining populations from those that are stable or increasing. Our inability to obtain this information has had important repercussions in the development of conservation measures to protect populations of bats.

Newborn bats grow rapidly to adult proportions, and it is usually impossible to distinguish between bats one and 10 years old. It is easy, however, to recognize a two- or four-month-old bat compared to an adult. Several techniques are commonly employed to determine the ages of mammals. Biologists will often remove and cut into sections a tooth. Like the trunks of some trees, the teeth of many mammals show annual rings, permitting observers to estimate with considerable accuracy the animal's age.

The problem of applying this method to bats is one of scale. In studies of bears and foxes, biologists usually extract one of the small premolar teeth located just behind the canine or eye tooth. This tooth is small enough to minimize discomfort and any disruption of the animal's life, yet large enough to section conveniently and examine in the laboratory. Even when these teeth are present in bats (Little Brown Bats have two pairs which would do; Big Browns have none), they are too small for easy removal and study. Extraction of a larger tooth, for example a canine, amounts to major surgery and could reduce the bat's chances of survival.

A possible, and less destructive, solution to the problem might be amputation of a toe; sections of the claw may provide a similar picture of annual rings. However, whether the rings are present in a tooth or a toe, proving that they are annual is another problem. The age of some mammals can be determined by examining the lens in the eye, but this is a post-mortem technique, of little use to biologists trying to obtain information about living animals.

It is often possible to distinguish between several degrees of tooth wear by examining the canine teeth of hand-held bats. The amount of wear is presumed to reflect the age of the individual, but studies of captive bats indicate tooth wear is influenced by the type of food consumed, and does not, in any event, provide absolute ages.

Most records of the ages of bats are derived from banding studies, although some are based on the years reached by animals in captivity. Two Little Brown Bats known to have survived at least 30 years in the wild were banded as adults at an old mine in Ontario in the late 1940s and were still alive when recaptured in the winter of 1979. However, the use of banding to determine age has important limitations. Failure to recapture an individual does not mean it has died; it could reflect its switching to another roost. Movement between roosts is common, and almost every biologist who has spent time banding bats is aware of the rapid rate at which banded (= disturbed) individuals disappear from the population.

Lest the interest in the ages of bats seem misplaced, it should be said that accurate assessment of the status of populations depends upon knowing not just how many animals are alive at any time, but also on being able to predict how many young will be produced the next breeding season. Estimating the population's capacity for reproduction depends upon information about age structure. Biologists concerned about the conservation of bats would benefit from information about age structures of their populations (see page 149).

What factors influence bat populations? Certainly, reproductive rates and life expectancy are two important elements affecting the populations of any animal. As we have seen, bats have low reproductive rates, but they live longer than most mammals their size. Many rodents, on the other hand, produce large numbers of young annually, but have relatively short lives. Although the two Little Brown Bats that had survived more than 30 years were probably exceptional, banding studies indicate that ages of 10 years are common. In her first 10 years, then, a female Little Brown Bat could produce 9 young, while a female White-footed Mouse that lived to the ripe old age of 3 years could produce 60.

We also know that because of the toll they take on young bats in temperate areas, hibernation and migration have considerable influence on some populations of bats. Populations may also be slightly affected by predators, and there is some evidence that populations may be limited by food supplies. There is no clear evidence of populations of bats being adversely affected by infestations of parasites.

Several studies have suggested that populations of some species of bats are limited by the availability of roosts, particularly roosts required for only part of the year. For instance, the distribution of Little Brown Bats is probably more a reflection of the location of suitable hibernation sites than the distribution of day roosts used in summer. Most Little Browns observed close to the northern treeline are males, while the ranges of females appear to be restricted more to the south, perhaps reflecting the different summer habits of the sexes. Females have limited time between their departure from hibernation and the birth of their young to reach suitable nurseries. Furthermore, much of that time is needed to replenish exhausted supplies of energy in preparation for the taxing demands of lactation. Because males do not face these onerous physiological demands between the time they leave hibernation and appear for mating in late summer, they have more freedom to wander.

In some cases, the density of bat populations is influenced by specialized roost requirements that severely limit available roost space. As we have seen (page 88), Spix's Disc-winged Bats that roost in unfurled leaves can only use these sites for one night. Consequently, a shortage of suitable leaves on any one night could result in a limit being imposed on the number of these bats which occur in one area.

Sometimes, the food supply in the area around a roost harbouring large concentratrions of bats is not sufficient to feed the entire population. In these cases, some of the population is forced to travel long distances to uncrowded areas where there is enough food. These feeding grounds are presumably not populated by the bats because of a lack of suitable roost sites, and the bats have little choice except to become commuters. Some Mexican Free-tailed Bats, for example, appear to make nightly round trips of more than 100 kilometres to and from areas of suitable feeding habitat. This, of course, raises the question of who gets to feed close to the roost. In some fruit-eating bats, it is the harem males (see page 127) that have this privilege.

Perhaps the strongest indication of the importance of roosts to populations of bats is the brief time required by many species to find and exploit new roosts. From the time of its opening in 1969, the National Arts Centre in Ottawa, Canada, was visited occasionally by bats, particularly in August and September. Within a few years, almost 500 bats were thought to visit the Centre to spend an occasional day in August and September (the bats used the place as a day roost). The bats usually entered and left the building through small cracks under the frames of two doors on the roof, although a few strays flew in through open doors, coming in via the parking garage or directly into the foyer area. It is remarkable not only that the bats discovered the inconspicuous openings on the roof, but also that so many used the Centre as a day roost. As a footnote, it should be added that the situation did create a few conflicts on evenings when performances were scheduled. The bats' departure for their evening's activities usually coincided with the arrival of patrons, and some bats took a wrong turn on their way out, adding considerably to the entertainment. We sealed the holes on the roof, which immediately reduced the bat traffic in the Centre and alleviated conflict between bats and people. (For more information about keeping bats out of buildings, see page 143).

The situation at the arts centre is another indication that bats continually search for suitable roosts. It also suggests that bats have an effective method of communication which permits them to exploit each other's finds. There is some evidence that in some species this communication may involve echolocation calls (see page 134).

Because bats are difficult to catch in some situations, their roosts are often unknown, difficult to find, or inaccessible, and since the question of the age structure in their populations remains a mystery, our understanding of the dynamics of their populations is largely incomplete. But despite these limitations, the information available about many populations of temperate bats in Europe and North America strongly suggests drastic declines in the numbers of some bats. A 1980 summary of bat population studies showed that during the past 25 years, seven of eight species known to hibernate in the Netherlands have been hit by marked declines in numbers; the eighth species appears to have increased in numbers. In North America, population plunges have prompted authorities to place at least two species on the Endangered List, giving them the same legal status as Whooping Cranes. Many of our efforts to conserve bats, however, are limited by lack of vital data about their populations. Even without answers to these questions about bat populations, there are real threats to the survival of some species which require immediate attention (see page 150).

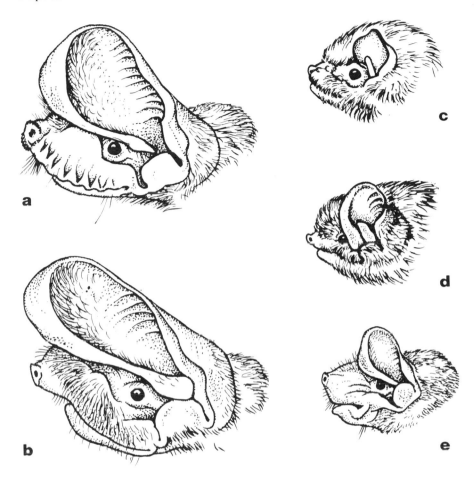

Big Free-tailed Bats (a) typically roost in cliffs in the southwestern United States, and their numbers can be assessed by counting them as they leave in the evening to feed. This process is facilitated by their distinctive echolocation calls, which are audible to the unaided human ear. Counting Thomas' Mastiff Bats (b) as they leave their roosts is more difficult as their echolocation calls are ultrasonic and huge numbers of these bats may occupy one roost. For different reasons, counting Common Bent-winged Bats (c) as they leave their roost sites is impractical: this species roosts alone in foliage. Population counts of Red Bats (d) leaving their roosts are not possible because we do not know where they roost. Predators may also wait at roost entrances to meet departing bats, and large colonies of Pallas' Mastiff Bat (e) have been exploited by some hawks and owls. Sketch by Paul Geraghty.

Predation and Mortality

When the diversity and large populations of bats are considered, it is surprising to learn that there are relatively few animals making bats a regular part of their diet. In fact, there is little evidence that any animal except man, through his destruction and poisoning of habitats, has much impact on populations of any species of bat. But this is not to say that many animals do not occasionally prey on bats. In Ontario's Algonquin Park, for example, a Pine Marten was observed using a nearby colony of Little Brown Bats as a source of food. In New Mexico, Great Horned Owls have been seen harvesting Mexican Free-tailed Bats from the swarms leaving Carlsbad Caverns. Other records show that domestic cats, snakes, raccoons, skunks, and many species of hawks and owls will also eat bats when they are available.

However, in most of these cases, predators were taking advantage of situations where large numbers of bats congregated in a predictable manner. The house cat observed waiting at the entrance to Wyandotte Cave in Indiana, catching bats as they came and went, was merely exploiting an easy, available food supply. In other situations, White-footed and Deer Mice inhabiting caves and old mines used by bats as hibernation sites occasionally supplemented their winter diet by eating dead or dying bats that had fallen from the walls or ceilings. The bats increased the mice's chances of surviving the winter by decreasing the amount of other food they had to find.

In the tropics and subtropics, where bats are more diverse and their populations larger, some birds regularly exploit them for food. Perhaps the most efficient of these predators is the Bat Hawk which is found in most of Africa and parts of the East Indies. One Bat Hawk in Zambia was observed catching free-tailed bats at the rate of 18 per hour. These hawks have large eyes and excellent powers of vision, and usually grab bats in mid-air with their talons, quickly stuffing them into their large mouths. They often hunt around the entrances to caves and other day roosts, although they sometimes adopt a more general searching mode in their foraging. They usually concentrate on bats weighing 30 grams or less (an adult Bat Hawk weighs about 650 grams) and may occasionally add some variety to their diet in the form of swifts, swallows, and insects. A trained Bat Hawk could be of great assistance to biologists trying to sample bats that fly high and fast; indeed, the first record of the Long-crested Free-tailed Bat was one that was removed from the stomach of a Bat Hawk which was shot in Uganda. Although several species of South American raptors sometimes include bats in their diet, none seems to specialize on them to the degree of the Bat Hawk. Despite their names, Bat Falcons from South and Central America feed most often on birds and insects.

Perhaps the most consistent enemies of bats in the tropics are other bats. In Central and South America, Peters' False Vampire Bats and Linnaeus' False Vampire Bats often include other bats in their diets. In Africa, Large Slit-faced Bats often eat other bats, and Heart-nosed Bats may do so occasionally. Two species of Indian False Vampire Bats from India and southeast Asia and Ghost Bats from Australia also prey on other bats. However, we do not know how important other bats are to the diets of these cannibalistic species, nor is it clear how they capture their bat prey, although there are some observations of Linnaeus' False Vampire Bat and Large Slit-faced Bats (see page 56) chasing and catching bats. We do know that the bat-eating species, which weigh from 40 to more than 100 grams, seem to attack bats weighing less than 20 grams.

The apparent dearth of predators specialized to feed on bats leads back to a question facing biologists studying the predator prey interactions of many animals. Is the absence of predators real? If we knew more about the potential enemies of bats would we discover that many animals do indeed rely on them as food? Or is the lack of predators a function of effective defensive behaviour, meaning that the lunar phobia exhibited by many bats throughout the world is in fact an anti-predator behaviour? Only more work will provide us with answers to some of these questions, and the answers are unlikely to be the same for all species of bats.

Accidents account for the deaths of many bats. Bats have been found impaled on barbed wire fences, thistles, cacti, and burdocks; in these cases the bats seem to have made a fatal miscalculation while flying. Although many of these accidents occur at times of the year when young are learning to fly, the fact that some occur at other times indicates that adults are not immune to such mishaps. One possible explanation for these casualties is that bats sometimes fly by memory, paying little attention to what is going on around them. It is this weakness which makes them vulnerable to trapping or capture in mist nets (see the Andrea Dorea Effect, page 13).

The fossil record shows that accidents are not restricted to errors in flight. Rising floodwaters have wiped out large numbers of bats hibernating in caves. In many parts of the world, heavy rains produce rapidly swelling floodwaters that could drown hibernating bats before they had time to arouse and flee to safer sites.

The inclination of bats to approach places from which others are calling also contributes to accidental death, and to their vulnerability to predators, nets, and traps. One observer in Salisbury, Vermont, reported finding 350 dead Little Brown Bats in a china slop jar that had been placed in the attic of an old hotel to catch water coming through the roof. I once found 100 dead bats in a cylindrical cage that had inadvertently been left behind in an old mine a few weeks earlier. The narrow neck of the slop jar and the smooth galvanized surface of the cage had evidently trapped bats inside; in either container the diameter of the neck was too small to permit the bats to fly out. It appeared that once one bat had ventured inside and become trapped, its calls attracted others.

Severe seasonal weather places many extra demands on bats, especially those in their first year, and is another major reason for death. Young bats lose weight after weaning, probably because of problems adjusting to hunting their own food. These same bats must then face the challenge of migration or hibernation, or both, activities requiring large amounts of energy. In some cases the young have not had enough time to put down the reserves of energy necessary for either operation and fail to survive. Although it appears that this situation sets the stage for widespread mortality, there are no records of mass deaths of young bats either during migration or hibernation, except in cases involving pesticide poisoning.

Like other animals, bats store pesticides in their body fat. Because their small fat reserves are largely used up while they adjust to hunting, young bats are more prone to pesticide poisoning; the concentrations of poisons increase as the store of fat declines. The mass deaths of Mexican Free-tailed Bats from pesticides during the 1970s at some sites in the American southwest mainly involved individuals born that year. In this case, it is likely that the lactating females had fed on insects sprayed with insecticides and then passed the contaminants along to their babies as components in their milk. Die-offs of Little Brown Bats in nursery colonies in New England have also been traced to pesticides, but in these cases the bats seem to have been in double jeopardy. While they probably picked up some poisons from their insect food, other insecticides (such as DDT) had been sprayed into their roosts to 'control' them.

The low reproductive rates of bats and records of impressively old age for some species are consistent with low rates of mortality after the first year. Few predators seem to specialize on bats, which leads us to believe that accidents and failure to survive the first winter are the main sources of natural mortality. But man is the main enemy of bats. Poisoning by pesticides, whether directly (in bat control) or indirectly (via the insects), is perhaps the most significant artificial cause of the premature deaths of bats.

Parasites

The relative scarcity of animals that make bats a regular part of their diet is more than balanced by the array of parasites that prefer to live in or on bats. This does not make bats unusual – most animals harbour similar levels of parasites nor does it mean that they are dirty. Indeed, bats are very clean animals, spending considerable time each day grooming their fur and licking their wing membranes. As a result, for a parasite to live successfully in the fur or on the wings of a bat, it must be designed to minimize its chances of being groomed off and eaten by its host. Similarly, a parasite living inside a bat must find a way to avoid being evicted by the immune system or, if it inhabits the digestive tract, being digested.

Parasites living inside animals are called 'endoparasites,' those living outside 'ectoparasites.' Bat endoparasites may include single-celled trypanosomes, tapeworms, flukes, and roundworms; among the ectoparasites may be fleas, bedbugs, mites, batflies, and other arthropods. However, biologists are still trying to catalogue the parasites of bats, and our knowledge about rates of infestation in different populations, their effects on their hosts, and the details of their life-cycles are not well known.

We do know that bats are the preferred hosts for many parasites, including bat mites, two families of batflies, and one family of bat fleas. The bedbugs that live on bats are distinct from the ones that live on people. Many other parasites, including some ticks, are not as fussy and will feed on bats or on other mammals. The fleas of bats are usually restricted to them, unlike the fleas of dogs and cats (as many know). Clearly, though, bats offer a wide range of opportunities for parasites. Most of these parasites can be placed into five broad categories: the single-celled Protozoa; flatworms, the Platyhelminthes; roundworms, the Nematoda; acanthocephalan worms, the Acanthocephala; and joint-legged animals, the Arthropoda. The first four groups are endoparasites; the arthropods tend to be ectoparasites.

The single-celled parasites usually live in the bloodstreams of their hosts. These include trypanosomes that may cause a variety of diseases, although we know little of their effects on the health of bats. Trypanosomes which prefer bats do not seem to be dangerous to humans, even though other trypanosomes may cause serious diseases such as sleeping sickness. Many of the other single-celled animals living in bats are known as blood parasites, and while similar forms lead to malaria in humans, we do not know how or if they affect the health of bats.

Flatworms specialized to feed on bats include two basic types: tapeworms, which live in the digestive tract, and flukes, which may live in many parts of the body, but often choose the liver or bladder. None of the flatworms that prefer bats is known to inhabit man, nor does it appear that any poses a problem in human health. Mind you, we do have our own array of tapeworms with which to cope.

Many species of roundworms live in bats, often in the digestive tract, the gall-bladder, or the urinary bladder. Again, roundworms of bats are different from the ones which exploit humans and do not appear to represent a threat to our health. A common human roundworm, the hookworm, causes a great deal of misery to many people.

Although there is only one record of an acanthocephalan worm being found in a bat (in a digestive tract), the list of arthropods which exploit bats is lengthy. Arthropods, ranging from mites through ticks, fleas, and flies to earwigs, may live in the fur, on the wing membrane, or just under the surface of the skin. Some of these ectoparasites remain with their host both inside and outside its roost, while others jump off before the bats depart.

How some parasites infect bats remains a mystery. When a parasite lives on the outside of its host and retains some mobility, it is easy to see how it could move from one bat to another in a crowded roost. But the mobility of ectoparasites that shed their legs and/or wings to burrow into the skin is more limited, and they must remain with one host until they die. However, these parasites have mobility during earlier stages of their lives, and no doubt can move about to find their lifelong host.

Explaining how the endoparasites that cannot survive outside their host travel from one bat to another is more challenging. In many cases the parasite relies on another animal, called a 'vector,' to move it from one host to another. For example, some bat trypanosomes are moved between bats by bedbugs which ingest the trypanosomes when they consume a meal of blood. Other one-celled parasites may be spread by the bites of mosquitoes. However, the life-cycles of flukes, tapeworms, and roundworms are often more complex. They seem to infect a bat by means of the food which it consumes – often insects harbouring different stages in the life-cycles of the parasite. Some plain-nosed bats living in wet areas have higher levels of infection by tapeworms than those living in drier climates, which suggests that aquatic insects may play particularly important roles as stopovers for the intermediate stages of tapeworms. A tremendous amount of detective work is needed to unravel the complex life-cycles of these parasites, and our knowledge of the details for most parasites of bats is incomplete.

In other mammals, a typical tapeworm life-cycle begins when the adult tapeworm produces and releases eggs in the digestive tract of its 'definitive host' (the one housing the adult parasite). The eggs leave the host in its feces and hatch into larvae if the feces land in water. The larvae swim and seek out and burrow into the tissue of an intermediate host, often an aquatic crustacean. When this intermediate host is eaten by fish, the larvae burrow into the muscle tissue of the fish and encyst there, changing into adults only when the fish is consumed by an appropriate definitive host. One tapeworm with this general plan of life-cycle may grow to lengths of 10 metres in human beings!

Some larval flukes infect their definitive hosts by burrowing under their skins and entering the circulatory system. We do not know if bat flukes use this approach, but lots of human-loving flukes do. 'Swimmer's itch,' a common summer malady in many parts of the world, is caused by a larval fluke burrowing under the wrong skin – a man's instead of a cow's. The larva cannot live in man, and is attacked and killed by the immune system, causing a small irritation at the site of invasion. Swimmer's itch is a minor discomfort.

In many tropical areas, particularly in parts of Africa and the West Indies, another variety of flukes use humans as their definitive hosts and infections of these visitors can cause a debilitating affliction known as bilharziasis or shistosomiasis. (The former is an English term, the latter an American one.) The key to solving this malady lies in working with the intermediate hosts, some species of snails which thrive in warm, slow-moving fresh water. There has been a dramatic increase in bilharziasis in the Nile River valley below the Aswan Dam which can be related to a general slowing in the speed of the river and a proliferation of snails in many newly available habitats.

It is obvious when we consider the damage caused in humans by heavy infestations of endoparasites that these visitors can have severe effect upon their hosts. From this base, it seems logical to suppose that bats may be adversely affected by heavy infestations of their parasites. We lack the details, but it is important to remember that the well-adapted parasite, particularly one living inside, does not kill or completely disable its host.

Because they can be observed alive and at work on the surface of bats, we know much more about ectoparasites. Mites and ticks often live on the fur or on the wings of their hosts, and their position on a bat's body often serves as a clue to the identity of the parasite. For example, some adult mites live only in the gums of Sanborn's Long-tongued Bat.

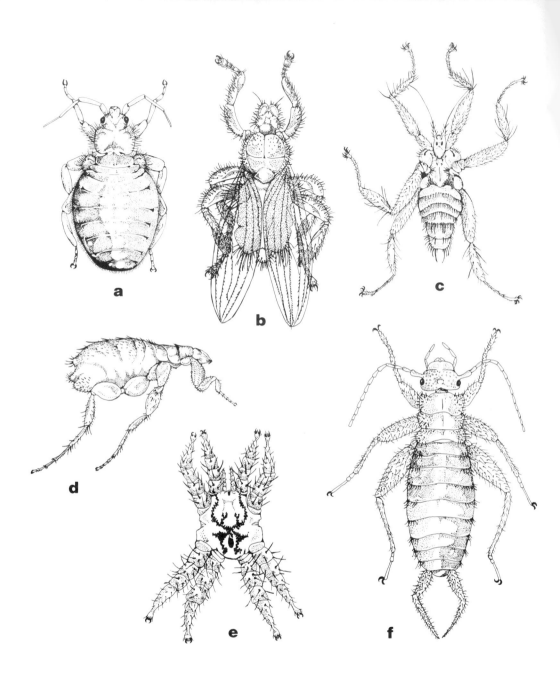

Some of the arthropods known to be ectoparasites of bats include bedbugs (a), batflies, either streblids (b) or nycteribiids (c), fleas (d), and mites (e). The earwig (f) that lives with Naked Bats is not a parasite. Sketch by Connie L. Gaudet.

Mites known as *Cryptonyssus robustipes* resemble tiny dots with eight legs when you see them running about on the wings of a bat. (Please excuse the scientific name, but 'common' names for the parasites of bats are not available.) The larvae of this mite do not feed, although slightly larger forms called 'protonymphs' and the adults consume blood or tissues. The smaller protonymphs pierce the skin of the wing membrane with their mouthparts and feed on tissue fluids. The larger protonymphs and adults consume blood which they obtain by rupturing small blood vessels and feeding on the subsequent haemorrhage. These mites do not insert their mouthparts into the blood vessels in the manner of most other blood-feeding arthropods.

Perhaps the most conspicuous of bat ectoparasites in North America are the trombiculid mites. During their immature stages, these bright orange mites often form small clusters on the ears and wing membranes, where they are thought to feed on blood. In some parts of the world trombiculid mites may be very common in the bat population; for example, about 15 per cent of 1,000 Little Brown Bats I surveyed in Ontario harboured these colourful ectoparasites.

Of the many insects considered parasites of bats, none is more interesting than two types of earwigs that live exclusively on Naked Bats, a large hairless free-tailed bat from Southeast Asia. Because they are the only known parasitic earwigs, an English zoologist, Adrian Marshall, spent some time unravelling their interactions with the Naked Bats. These earwigs lack the specialized mouthparts for piercing skin and sucking blood or tissue fluids that are typical of most ectoparasites. Careful observation showed that the earwigs live on secretions of the bats' skin, bits of dead skin, and exudates from the ears, eyes, mouth, anus, and genitals. When the bats were absent from their roost, the earwigs could survive for short periods by feeding on the bats' droppings. Marshall's study, then, showed that these earwigs are not parasitic at all, just hangers-on helping themselves to some bat by-products.

However, most of the insects tailored to live with bats collect meals of blood from their hosts, and probably do no favours for the bats. Some of these insects have developed unique specializations permitting them to exploit the opportunities which are offered by bats. Excellent examples are provided by the batflies, the Streblidae and the Nycteribiidae, two families known only as parasites of bats and most common in the tropics.

These flies have modified their life-cycles in ways that make them more adaptable to living with bats. The life-cycles of most flies have four separate stages: egg, larva, pupa, and adult. During the egg stage, a single cell changes into a multicellular organism that hatches as a larva. Insect larvae, familiar as caterpillars or maggots, are usually 'feeding machines' that eventually enter a dormant phase, the pupa, in which they change (metamorphose) into the adult stage; adult insects are usually 'mating machines.'

A Western Big-eared Bat (10 grams) shows the enormous ears typical of some insectivorous species. These bats are common in parts of western North America where they roost in caves and mines, and occasionally in buildings. During hibernation the streblid flies which parasitize them do not reproduce.

Batflies give birth to 'live young.' To achieve this, the females retain the eggs inside their bodies, where they hatch and pass through most of their larval stage. At birth, then, the young are not fully developed as pupae, and the female must find a secure place for them to pass through their pupal stage. Female nycteribiids that live on two species of Club-footed Bats in southeast Asia usually deposit the pupae on the tops of the bamboo nodes in which the bats live (see page 83). By depositing them on the ceiling of these mini-roosts, the female fly ensures that the pupae will be out of the bats' way, but still relatively close to the bats, whose presence has a direct effect upon the duration of the pupal stage.

When bats are in residence, the pupal period lasts about 25 days; in their absence, it may go on for up to 50 days. In the laboratory, three conditions will induce emergence from the pupal stage: when bats are placed in the container with the pupae; when someone breathes on the pupae; or when their container is shaken. By emerging only in response to stimuli that would be associated with the bats (carbon dioxide and movement), the flies have a built-in safety factor ensuring that the newly emerged adults have an immediate blood meal.

Trichobius corynorhini, a streblid fly commonly found on Western Big-eared Bats in North America, has developed a similar life-cycle, although it is modified to permit survival under cooler conditions. While tropical streblids and nycteribiids are able to reproduce throughout the year, the *T. corynorhini* flies living in regions with prolonged cold conditions – for example, a study site in Kansas – do not breed in the winter. The flies spend the winter with the hibernating bats, usually choosing female bats as their winter hosts. Female Western Big-eared Bats congregate in nurseries in summer, thereby increasing the flies' chances of encountering more bats.

In the end, it must be said that the specializations developed by the parasites of bats are as fascinating as those of their hosts, and provide us with more textbook examples of features designed to allow an animal to fully exploit its environment. Sometimes, these specializations offer biologists more than just a better understanding of life. Looking at parasites is often an easier way of identifying similar species of bats than examining the bats themselves. In parts of Australia, for example, the Little Pipistrelle can be distinguished from the Australian Little Brown Bat because it plays host to a hairy bat fly (*Basilia dispar*), while its colleague is inhabited by a bat fly with bristles (*Basilia musgravei*).

Behaviour

Many times when people learn of my fascination with bats I am asked: 'Just how smart are they?' It is difficult to measure accurately the intelligence of humans, let alone bats, so I usually answer with a favourite anecdote.

To get an indication of how well they respond to changing conditions, I decided to attempt to catch Little Brown Bats as they emerged from a day roost in a crack between two sections of a building where a new wing had been added. For several evenings I had noticed the bats leaving to feed by launching themselves out from the building and then flying away. From this information I attached one end of a mist net to the main part of the building and extended the remainder out parallel to the wing so that it was about half a metre away from the wall of the new wing. My strategy was successful, and I nabbed more than 100 females on first night of netting. I returned a week later and set the net in the same way. This time, however, I caught ony 15 bats; most of them avoided the net by going under it or waiting in the roost until I had taken it down. By the third week, I was only able to catch two bats at this site, and most of the colony exited by stepping out of the entance to their roost and, keeping their wings folded, dropping straight down until they were below the level of the net, and then taking flight. The response of these bats to the setting of the net was most impressive, an indication that Little Brown Bats, at least, have the ability to learn and remember.

But there is another side to this glowing picture of bat intelligence. During a visit to the Lamto Ecology Station in the Ivory Coast, I set out to capture some insectivorous bats by placing mist nets over a small bayou near the Bandama River. After several nights the nets had caught no insectivorous species, but they regularly caught Buettikofer's Fruit Bats active in the area. One night stands out in my memory. One of the nets was set across the bayou so that I could stand on a fallen tree to remove anything that flew into it. That evening when I went to check the nets, I found the bank of the bayou and the fallen tree carpeted with a colony of aggressive army ants. To get to the net, I had no choice but to walk through and then stand in the teeming mass of ants. The first four bats in the net presented no serious problems. I untangled and released them without incident, although the ants took offence at my presence and launched numerous effective attacks inside my trousers.

The fifth bat spelled trouble. I spent three long minutes untangling him, making occasional, ineffective attempts to brush off the ants that were now swarming up the inside and outside of my pant-legs. I was finally able to free the bat, but he immediately flew directly back into the net. He was upset; I was becoming frantic. Five times we repeated the exercise, and each time I worked to untangle him he screamed in my ear, setting a tempo that seemed to encourage the ants. The sixth time the bat finally realized his grief (and mine) would end if he stayed out of the net.

The American zoologist Jack Bradbury, who has studied Hammer-headed Bats, relatives of Buettikofer's Fruit Bat, had suggested that these bats were relatively feeble-minded. Only half-jokingly, he proposed that their brains showed just three responses, flying, feeding, and mating. He did acknowledge that screaming might be added to the repertoire. The behaviour of the bat entangled in my net did nothing to make me doubt Jack's view of the bats' mental capacity. However, it might be suggested that the bat, annoyed at being ensnared, was merely exhibiting a perverse sense of humour and deliberately exposed me to the maximum amount of discomfort.

These contrasting anecdotes seem to imply that some bats are better able to respond to adverse conditions than others, and that insectivorous species may learn more quickly than fruit-eaters. Perhaps that is so, but the behavioural responses of bats cannot be well assessed by their reactions to traps or nets; indeed, their ability to avoid capture is often impressive. To some degree, the behavioural patterns of bats are governed by reflexes acquired through experience. In this regard they resemble people who look both ways before they cross the street. A visitor from another planet might suspect that this street-crossing reflex was inherited if he were to observe pedestrians waiting to cross a busy street. But we know that this response is flexible and gained through experience, as any North American visiting Britain, where cars drive on the left-hand side of the road, quickly realizes.

Other reflexes, however, do not have to be learned; you do not need to be told or to learn to remove your hand quickly from a hot stove element. Bats also have an arsenal of these 'hard-wired' or automatic responses. One of them appears to be grooming, the daily ritual of combing the fur with the claws of the hind feet, and carefully licking the wings and other membranes. When combing, bats move their hind feet up and down vigorously, and by impressive contortions are able to reach all of their fur. Material combed out of the fur is licked from the claws before the process is repeated. Interestingly, the lower incisor teeth of some free-tailed bats are specialized for use as combs. Bats are able to remove dirt from both sides of their wings by moving them into unusual

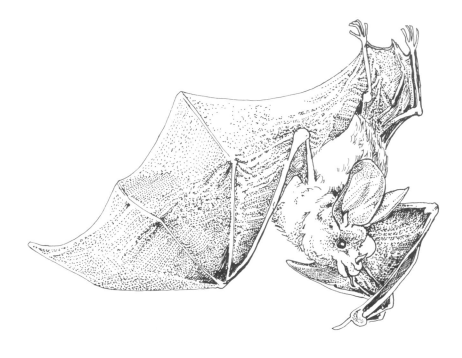

positions and giving them a thorough going-over with their tongues. The attention paid to cleansing any part of the body depends upon the amount of dirt that has collected there. After spilling some dog-food on part of one wing and its chin, a pet Hoary Bat spent three times longer than usual cleaning the soiled areas.

Several aspects of the behaviour of bats emphasize their diversity, but our knowledge of their importance is meagre considering the number of species available for study and the few for which there are data. However, considerable emphasis has been placed on studies of the sex lives of bats, providing us with a useful jumping-off point for a consideration of their behaviour.

SEX

Because they produce milk, the females of most mammals, including bats, rear their young with little assistance from the males. A few mammals have only one mate during a breeding season and still fewer are monogamous for several seasons or for life. Among many of these monogamous species, males assist the

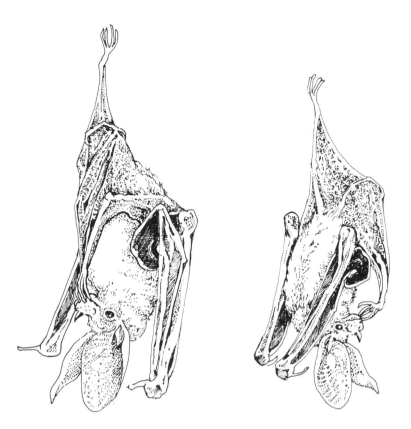

This California Leaf-nosed Bat shows grooming
movements typical of bats in general. The tongue is
used to clean the wing membranes, and the claws on
the hind feet to comb out the fur. The bat uses its
tongue and teeth to clean its comb. Sketch by Connie
L. Gaudet.

females with the care and rearing of the young, perhaps by bringing her food which she converts into milk. However, monogamy is exceptional among the more than 5,000 species of mammals. Furthermore, only some of the males of many mammals are reproductively active. In these situations, the females have some means of assessing a male's potential to sire young, and a male's behaviour often provides evidence of a mating system functioning in this manner. Males may engage in real or mock fights with other males to demonstrate their physical prowess, or they may stage 'displays' to convince passing females that they are the best choice. In either case, a female's involvement with the male she selects is brief, often only the time it takes to mate. Having obtained what she needs, the female then sets out on her own to await the birth of her young. These mating systems are called polygamous, and may involve one male with several females (polygyny), or one female with several males (polyandry). When there is no pair bond between the male and female, the males and females may consort with several mates, the system is called promiscuous.

People living in parts of the African rain forest have long been familiar with a nocturnal cacophony usually staged along the edges of rivers by assemblies of male Hammer-headed Bats. At the beginning of each rainy season in Gabon (July and August, and January and February), males congregate at their calling sites after dark, usually spaced at intervals of about 50 metres in trees along the edge of the river. The males hang from branches and flap their wings, increasing the rate of their 'honking' as each female flies past. An observer equipped with this information can walk up and down the line of calling males and assess their numbers, as well as the number of times females pass the calling sites. Furthermore, since the females give a characteristic groan-like call after mating, it is possible to score the number of successes of the males in the assemblage. When this tally is made, it is obvious that males calling from the middle of the line mate more often than those at the ends.

Biologists call this type of aggregation and mating system a 'lek.' This term is appropriate when the males are the only resource at the display site, when there is a marked difference (dimorphism) between males and females, and when some males are more successful than others. Hammer-headed Bats have mating systems which meet all of these conditions; the sexes are strikingly different, the females feed and roost away from the display sites, and some males are more successful than others. By attaching small radio transmitters to the females, biologists have discovered that they often make several passes up and down the line of calling males before choosing one as a mate. Observations of marked males revealed that the central or most active positions in the array were occupied by the same individuals night after night, a convincing indication that some males are indeed more successful than others. It is not clear what criteria females use to choose their mates, although the honking and wing-flapping

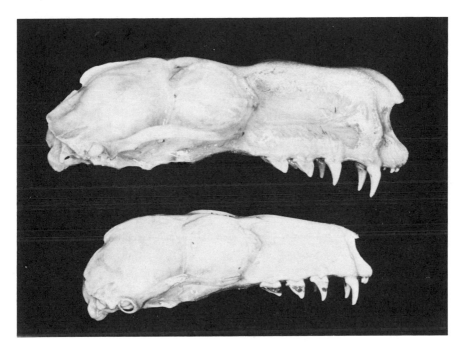

The skulls of a male (72 millimetres long) and female (62 millimetres long) Hammer-headed Bat provide a striking demonstration of sexual dimorphism associated with lek behaviour. The inflated rostrum or snout area of the male appears to act as a resonance chamber contributing to the production of sound which he uses to attract females.

probably exert some influence on their selection. As far as we know, mating does not occur away from the leks, and females that have made their choice and mated are not harassed by other males.

The lek system stands in sharp contrast to other mating systems. In species forming harems, a situation where one male lives with several females, there appears to be a strong pair bond between the bats. Among many species of New World Leaf-nosed bats, single males live for more than a year with groups of up to 30 females. Genetic studies of Spear-nosed Bats reveal that the male usually sires more than 95 per cent of the offspring born in his harem. However, the social structure of this species is poorly understood. Beyond mating, the male seems to have little to do with his females, although he guards them in their day roost, keeping other males at bay. If removed he is quickly replaced by one of the bachelors loitering around the cave. The females play no obvious role in the

During mating, the male Little Brown Bat approaches the female from the rear and grips the scruff of her neck in his teeth while holding her with his thumbs. These common North American bats mate in the late summer and autumn, and the females store fertile sperm through the winter.

selection of a successor to the harem male. The females in the harem are not close relatives, sisters or half-sisters, but they are usually the same age, and form a cohesive social unit, getting back together if removed from their day roost, and apparently foraging together as a group.

The females of some species change harems more often. Female Greater White-lined Bats may select a different harem each day, prompting the males to stage lively displays at dawn to win their favour and attention. The males mark the boundaries of their roosting territories with secretions shaken from their wing glands, and perform flight displays to identify the associated airspace. In a related species, the Lesser-White-lined Bats, the males control feeding as well as roosting territories, and members of a male's harem can use his roosting and feeding territories without interference. Under these conditions, females may select a male on the basis of his physical attributes, or those of his roost or his feeding territory.

Other species of bats form harems that mix stability and flexibility. For example, female Short-tailed Fruit Bats usually remain in one male's harem for longer periods than the female Greater White-lined Bats, although they occasionally select a new male and move to his harem. As we shall see, these exchanges and the interactions of females with the harem male and other females have important repercussions for young bats.

Little Brown Bats exhibit a more chaotic pattern of male-female interactions. At first glance, I thought that their mating system would provide another example of a lek. The males perched at positions along the walls of caves and mines serving as mating sites, and there produced echolocation calls. The females flew up and down the passages, each eventually joining a male, and, in some cases, mating with him. But more careful observation showed that while males mated more than once, often with different females, none seemed to be more sexually successful than others. This ruled out the possibility that the mating system was a lek.

Furthermore, as the season progressed, the mating behaviour of Little Brown Bats changed along with a general decline in activity. Mating from mid-August to early September involves active bats, but by mid-September more bats enter torpor. As this happened, the males changed their strategy, moving from one cluster of torpid bats to another, selecting individuals and attempting to copulate with them. 'Attempting' is the only word to use in this case; the bats accosted were often other males, and partners of either sex usually did not arouse from their winter sleep during the encounter. In the active phase of mating, males used distinctive copulation calls to calm struggling females objecting to being mounted.

The mating system of Little Brown Bats is clearly promiscuous. There is no evidence of a pair bond between male and female; males mate with more than

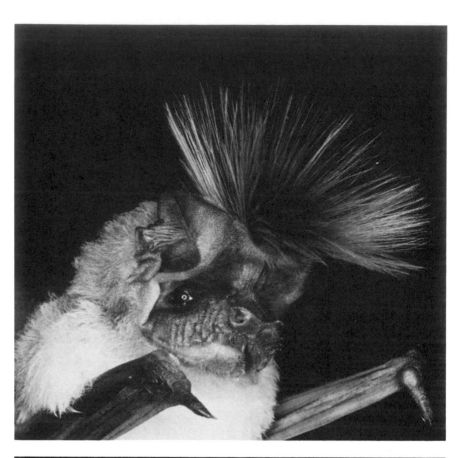

one female, females with more than one male, and males have been observed trying to mate with other males. The system does seem rather chaotic, and I have lightly referred to it as a 'disco' mating system. But why has this apparently random an chaotic system evolved? In the lek and harem systems, there is strict order and the choices of females are respected in most cases. A possible explanation for the mating system of Little Brown Bats involves hibernation. When they become torpid, the males have no means of keeping other males from their females; for the same reason, females have no obvious way to protect themselves from unwanted copulations. In other words, neither males nor females can protect any investment of time and energy they make in choosing a mate. In this situation you would expect males to mate with as many females as possible, and *vice versa*. Another intriguing possibility, much more difficult to prove, is that females allow any male to mate with them, relying on some mechanism to ensure that only the best sperm survive the winter to fertilize their eggs in spring.

As we have seen, female horseshoe and plain-nosed bats store sperm through the winter (page 104), and some of these bats produce plugs to block the vagina. It is possible that female Noctules and some horseshoe bats – for example, Greater Horseshoe Bats – use vaginal plugs as 'chastity belts' to protect their investment in mate choice. However, as noted previously, the case for vaginal plugs as blocks to prevent loss of sperm is equally convincing.

Certainly, what we know about the sexual behaviour of bats does not make them sterling examples of loyalty or tranquil domesticity. However, the sex lives of only 20 species of bats have been examined in any detail, and the variations evident from these studies imply that biologists can look forward to many interesting surprises as they broaden their data base. Personally, I am anxious to learn about the mating systems of Big Brown Bats, an extremely common species whose sexual behaviour remains virtually unexamined. I also look forward to answers about the role of the long, erectile head crests of male Long-crested Free-tailed Bats in mate selection by the females.

(Opposite) Male and female Long-crested Free-tailed Bats (10 grams) provide a striking example of sexual dimorphism; the crest is prominent only in the males, and presumably plays some role in mate selection.

MOTHER-YOUNG INTERACTIONS

Our knowledge of the relationships between female bats and their young is just enough to whet the appetite. This interaction changes as the young grow older, and raises an interesting set of questions. At weaning, for example, do mothers bring solid food back to their young, perhaps regurgitating semi-digested materials in the manner of many other mammals, or are the youngsters on their own almost immediately?

When they are old enough to fly, young Heart-nosed Bats follow their mothers around, often accompanying them on hunting expeditions. Several times I have caught pairs of female Egyptian Fruit Bats and their young near the trees in which they fed, indicating that the mothers were accompanied on these foraging flights by their young. This assumption was supported by the behaviour of a young male I captured and held overnight with an adult female trapped in the same place at the same time. I placed the bats in separate bags, but during the night the young male escaped from his sack and by morning he was hanging on the outside of the bag containing the female. The fact that the female was not lactating suggests that even after weaning the young bat stayed close to his mother.

Of course, the only way we can expect to unravel the mysteries of mother-young interactions at weaning is by observing bats in their natural settings. However, this is often easier said than done. Despite extensive observations of Little Brown Bats in their nursery colonies, the relationship between mothers and young at weaning is far from clear. Some studies have suggested that females provide their young with insects, but the question remains unresolved.

Just as fascinating are interactions between mothers and baby bats. Species that do not form communal nurseries, for example Red and Hoary Bats, face one set of problems, while those gathering in huge colonies, such as Mexican Free-tailed Bats, are confronted by additional complications. The solitary species must remember the location of their roost, but bats from large nurseries must recall both the location of their roost and the identity of their own young. Greater White-lined Bats, which leave their young in secluded roosts while they hunt, must be able to find their sites each night.

To help their mothers locate them, the young of many species, including Little Brown Bats, produce signals known as isolation calls. For their part, mothers utter specific calls when approaching the young. Little Brown Bats usually establish medium-sized nursery colonies, although they often contain more than 1,000 females. An early experiment with captive Little Brown Bats suggested that mothers would nurse any young. This was demonstrated by providing the young with access to different females, including their own mothers. The young continued to grow when not confined with their mothers, which implies that the bats fed more than their own offspring.

However, later studies in the wild by Christine Thomson, when she was a graduate student at Carleton University, showed that mothers consistently recognized the isolation calls of their own young, and were very selective about retrieving them. The behaviour of mothers trying to collect their young suggested they first used sound and then odour to locate and recognize them. This study also revealed that while the mothers were selective about which young they suckled, the babies tried to obtain milk from any female. In many cases, babies trying to obtain milk from females other than their own mothers were repulsed, often with considerable force.

In one experiment designed to test the ability of females to recognize their young, a mother Little Brown Bat suckling her baby was presented with a choice of two loudspeakers, one carrying her baby's isolation calls, the other a control signal. The female cocked her ears at each speaker, looked under her wing and nudged her baby, and ignored both stimuli. Clearly she had recognized the calls of her baby, but was not about to search for something that was not missing.

Other strong evidence that females recognize their own young comes from work with captive Short-tailed Fruit Bats. Fran Porter, an American psychologist working at Washington University, found that her captive bats established harems each with one adult male and several females. In captivity it was possible to observe at close range the behaviour of the bats, and, since the animals carried diagnostic bands, to document the movements of individual bats between different harems. If a female dropped her baby, her harem mates (other females) and the male, usually the father, harassed her until she retrieved it. In this situation the bats identified specific babies with specific females, and responded accordingly when a youngster was in danger. The best example of this recognition involved a female that had mated with one harem male and then changed groups. When she dropped her baby, she was scolded by the other females in her harem, and by the first harem male, probably the father.

The available evidence suggests that bats, like other mammals, recognize their own offspring and take great pains to nurse only them. This selectivity makes sense considering that a female must invest a great deal of energy in rearing her young, and that milk given to another offspring is not available for her own. Again, however, we have data for very few species, and the full spectrum of mother-young interactios will only become apparent with further study. For example, preliminary studies of Mexican Free-tailed Bats and some bent-winged bats in Australia suggest that females in crowded nurseries form a 'milk herd,' suckling any young that approach them. But this work is far from conclusive, and more studies, particularly those involving individually recognizable mothers and young, are required to substantiate the earlier reports.

Biologists have tended to consider the echolocation calls of bats separately from their other vocalizations. But as our knowledge of bat behaviour grows, we are beginning to realize that the distinction is not as marked as we had thought. Indeed, many bat sounds may serve more than one purpose. We know that echolocation calls are crucial to successful orientation for many bats, but recent experiments with Little Brown Bats demonstrate clearly that the signals of individuals are often exploited by others. Robert Barclay, a Canadian zoologist, demonstrated that Little Brown Bats emerging from their roosts to feed at night would approach a loudspeaker emitting the echolocation sounds of Little Brown Bats and other species. To be certain the response was not a general sensitivity to sounds, Robert used other sounds as controls. The bats only responded to the echolocation calls or to artificial echolocation calls. When bats feed on insects which accumulate in swarms in small areas, there is a clear advantage to being able to find these patches of food quickly. Using echolocation, Little Brown Bats probably detect targets only when they are within 2 metres, but they can probably hear the echolocation calls of other bats at least 50 metres away, an indication of the advantage of eavesdropping on your neighbours.

Little Brown Bats also use the echolocation calls of others to locate day and night roosts. In this case, the pirating of signals is a two-way street; the eavesdropping bat finds a suitable roost, and the echolocating bat gets a roost-mate to help him (or her) stay warm. Experiments with Spotted Bats provide another example of how the echolocation calls of one individual may be used by another. When Marty Leonard, then a graduate student at Carleton University, played their echolocation calls back to them, the Spotted Bats responded by flying at or away from the speaker. However, they showed no response when other sounds, including their echolocation calls played backwards, were presented. Spotted Bats usually feed alone, and their response to presentations of echolocation calls suggested that they eavesdrop to avoid one another.

There can be disadvantages to having other bats eavesdrop on your echolocation calls. A bat trying to avoid detection while it feeds in another's territory immediately advertises its behaviour to the owner. Of course, the territorial bat benefits from being able to distinguish a poacher from a commuter; one may steal his food while the other is just passing through.

A purist would not agree that echolocation calls can be considered communication signals in all of these cases. By definition, some feel, there must be the intent to communicate on the part of the animal producing the signal. Thus, an echolocating bat approaching a night roost may be communicating if it is attempting to attract others to ensure warmth. On the other hand, the Little Brown Bat that attracts others to a rich patch of food might have preferred to keep the windfall for itself, and its 'communication' is not intentional.

Studies of the echolocation calls of Mexican Bulldog Bats hunting for fish revealed that these signals may be slightly modified to add a communication element. Rod Suthers, an American zoologist, found that a bat on a collision course with another lowered the frequency of its orientation sounds, producing a 'honk-like' noise that led one of the bats to veer away, avoiding a mid-air collision. Several other species are now known to use similar honks, and they must be considered communication signals, even by a purist's definition.

Bats produce a wide range of other vocalizations that probably serve no purpose in echolocation but play important roles in social interactions. For example, the copulation calls of male Little Brown Bats and the mother-young signals reported from many species serve specific functions. The many squawks and growls produced by most bats probably carry some social significance, and subtle changes in these sounds may have the same importance to bat interactions as changes in facial expression or posture have in encounters between people. The honking of a male Hammer-headed Bat advertises its presence and virility to passing females. Heart-nosed Bats 'sing' as they move about their feeding areas, apparently informing others of their presence and claim to the feeding rights. Pallid Bats produce a 'directive' call to attract others, often just before dawn as they are selecting a day roost.

From their glands bats exude a range of secretions with important social purposes, including the marking of females, young, and territories. How much of the recognition of young by females, or harem mates by one another, is a function of odour is not yet clear. And, as we have seen, visual displays are also used to communicate information, mainly about territorial boundaries and sexual prowess.

Communication between bats seems to involve any of the media to which they have access. But the importance and relative roles of these different media (sound, sight, smell) still need to be resolved in most cases. Sound probably offers bats good long-range communication possibilities, as can vision when there is enough light. Smell may be most effective at short range. Common sense dictates that much of the communication between bats will involve some combinations of the different senses.

SWARMING

Although it is convenient to treat different aspects of behaviour under separate headings, this rarely reflects the situation in nature. In the preceding pages, for example, I have presented information about mating systems and communication, but the one is vital to the other and their separation is arbitrary. The swarming of temperate bats at hibernation sites in autumn is an excellent example of a behaviour that encompasses many other activities, including interactions between adults and young, mating, and migration.

Swarming starts in midsummer and continues until the weather limits the insect supply and precludes outside activity. In eastern North America, bats arrive at hibernation sites, which are the centres for swarming, after they have fed, usually about an hour after dark. The build-up in activity at the swarming sites through the summer coincides with dispersal from the nursery colonies, and many of the bats arriving at the sites are young born that year. The ratio of these sub-adults to adult females is more than 1:1, indicating that the movement to the hibernacula does not involve all of the adult females (if it did, the ratio would be 1:1, since each female has one young). Bats usually arrive at the sites in groups, suggesting that some may follow others. On the rare occasion when one catches what appears to have been a group, it usually contains an adult male, some adult females, and some sub-adults of both sexes.

Most of our knowledge about swarming is based on studies of Little Brown Bats. During the first two weeks in August, there is a continuous increase in the populations at swarming sites, but banding studies have revealed that the population is usually composed of a different group of bats each night. The swarming bats, of course, do not remain at the hibernation sites through the day. By mid-August, some bats begin to mate, attracting the attention of others still flying around in the cave or mine. If a copulating pair has not found a secure roost, a small horizontal shelf in a cave, or a drill hole or ledge in a mine, they risk being dislodged by other curious bats clambering around and under them.

Along with the cooler weather at the end of August and early September come the hibernating bats, increasing in numbers as the autumn progresses. At this point, only a few of the bats at the swarming sites were there earlier in the summer, and only about 15 per cent of the bats which swarmed at a site will overwinter there. The disappearance of many bats from the population suggests that swarming is not just a mating phenomenon; it seems to be intimately involved with preparation for migration, and may also be a means of leading young to suitable hibernation sites.

Our understanding of the behaviour of bats is still in its infancy. Learning the full significance of swarming, for example, may help us to understand more about the mating systems and the movements of bats from summer to winter roosts. Studies designed to measure the ability of bats to respond to changes are constantly providing clues that we have only begun to appreciate their rich behavioural repertoire.

Experiments conducted in 1981 and 1982 by Connie Gaudet, a Canadian

These Pallid Bats (20 grams) appear to be in deep discussion about the absence of a food reward on the target to which they have flown. The split celluloid rings or bands permit an observer to tell one bat from another, and thus to compare one bat's skill at learning with that of others. Pallid Bats occur in western North America.

zoologist, provide an excellent example. In an effort to compare the behavioural flexibility of Little Brown, Big Brown, and Pallid Bats, she planned to train them to solve a problem, and then find out how quickly they could adapt to new problems. This approach, namely challenging the ability of bats, has provided significant advances in the study of echolocation, largely through the work of Jim Simmons, an American psychologist, and his colleagues.

As expected, Connie found that Little Brown and Big Brown Bats learned quite quickly; this reflected the experiences of other experimenters. The Pallid Bats, however, proved more difficult, and required almost six weeks before they performed properly. Performing properly meant flying across a room to a target and being rewarded there with a piece of food. The real excitement came later, however, when she put one of the trained Pallid Bats in with several untrained ones. The novices learned the task within two days, apparently by observing the trained bats, showing quite clearly that Pallid Bats are better trainers of Pallid Bats than are people.

These results raise many exciting possibilities. We know that some birds will learn by watching others, and the same is true of some mammals. We now know that bats will learn from other bats. The capacity to learn from each other has important implications for many aspects of bat behaviour, from the training of young bats by their mothers to the exploitation of new food supplies and roost sites.

Their rich repertoire makes the behaviour of bats an excellent reflection of their diversity!

Public Health

The popular misconception that bats are dirty and dangerous has led to feelings of distrust and hostility in many parts of the world. True, bats are commonly associated with two serious diseases, rabies and histoplasmosis, but their link to either disorder is not uniform throughout the world. Bats can spread these afflictions, and perhaps others yet to be discovered, to people, their livestock, or their pets, but except in some tropical regions there is no reason to treat them as 'threats to public health.' When dealing with bats, it is important to distinguish between things we do not like and things posing a real threat to our well-being. Indeed, by 1969 on a world-wide basis, there were only six confirmed human deaths attributable to rabies spread by the bites of bats.

RABIES

A disease of the central nervous system, the brain, and the spinal cord, rabies is caused by a bullet-shaped virus. It appears to be confined to mammals, although there is some evidence that birds may also be susceptible to it. The symptoms of rabies for bats, as for other mammals, include general malaise, restlessness, and paralysis, especially of the hind limbs and throat muscles. Paralysis of the throat muscles prevents swallowing, resulting in an accumulation of saliva that gives the impression the victim is frothing at the mouth. Spasms in the throat muscles are aggravated by attempts to drink water, hence the name 'hydrophobia.' With increasing paralysis, of course, the victim becomes less mobile; some biologists suspect that a bat has less than three days to move about when the infection reaches this stage.

Biting is the most common way of transmitting rabies from one animal to another. The virus is often concentrated in the saliva of the rabid animal, which usually ensures its entry into the bloodstream of the bite victim. Rabies can also be introduced into an animal through ingestion of infected food, although experiments with laboratory mice indicate the chances of transmission in this way are small. Some biologists have suggested that rabies virus can be transmitted by the bites of ectoparasites, but this remains to be conclusively demonstrated for the ectoparasites of bats.

A more frightening mode of transmission is known as the 'aerosol route.' Two people who entered Frio Cave in Texas in the 1950s contracted and died of rabies even though neither was bitten by any of the resident bats or their ectoparasites. Subsequent experiments with different mammals showed that exposure to the air in the cave could lead to infection by the virus. The air is strongly affected by the tens of thousands of Mexican Free-tailed Bats roosting there, and the atmosphere is humid and oppressive. For aerosol transmission to occur, the virus must somehow be suspended in the air, presumably in water vapour. Most likely, the virus gets into the air from the urine of infected bats through evaporation, and then into the lungs of other animals breathing the contaminated air. For this system to work, high temperatures and humidities and large populations of bats are probably essential. We do not really understand how this system works, but I, for one, would not be in any hurry to work in a large cave colony of Mexican Free-tailed Bats.

Because they feed on blood and are warm-blooded, vampire bats are ideally suited to pick up rabies virus from infected animals and spread it to others. The fact that they must feed daily ensures that the virus is transferred from one host to another, while the bat's warm blood guarantees the virus a warm place to live and prosper in the interim. Some individual vampire bats may be 'carriers,' which means they could harbour the active virus without showing clinical symptoms of the disease. Most other bats, and most vampires, are not carriers, and, if infected, will show the classic symptoms and succumb to the disease. It is important at this stage to remember that only three of the world's 850+ species of bats are vampires, and that they are confined to South and Central America. Rabies among bats becomes less common as you move away from the range of vampire bats.

It is difficult to understand how bats are involved in the cycle of rabies in other animals – for example, foxes, skunks, and jackals. Presumably a bat bitten by one of these predators would not be vulnerable to infection by rabies since the bite itself would probably be fatal. In the same way, transmission of rabies to humans by the bites of grizzly bears, lions, or tigers is unlikely, considering the damaging effects of the bites. It is possible that a skunk or fox could become infected by eating a rabid bat, or by being bitten by a rabid bat it was about to eat; however, experiments conducted in the United States suggest that this route of infection is unlikely.

How rabies is passed from bat to bat, members of the same or different species, is much easier to understand. Animals roosting together may transmit diseases, particularly if interactions in the roosts involve altercations including biting. In crowded, hot, and humid roosts, aerosol transmission might also be possible. Until recently, it was less clear how bats away from their roosts could infect one another with rabies. There were reports of foraging bats attacking other bats, but these were usually interpreted as territorial disagreements or attempts at cannibalism. Then, in 1978, Gary Bell was working at a site in southern Arizona and, in the space of 20 minutes, watched a Hoary Bat attack, bite, and drive to the ground a Silver-haired Bat, a Mexican Free-tailed Bat, and a Big Brown Bat. The Hoary Bat was rabid, and could have exposed three other bats (three other species) to the virus in a very short time.

Once inside a mammal, the rabies virus passes an incubation period of three weeks to several months. But stress can accelerate activation of the virus and appearance of the symptoms of rabies. Experiments with hamsters have shown that crowding will activate the virus, and that injection of hormones that simulate the effect of crowding have the same effect. It is likely that the stress induced by pesticide poisoning could also activate dormant rabies virus.

As a rule, rabies is separated into two clinical forms: 'furious' and 'dumb' rabies. An animal with the furious form usually goes berserk, attacking anything and everything in its path. Dumb rabies is characterized by increasing paralysis that progressively weakens and immobilizes the victim. It is possible that an animal with rabies could go through both furious and dumb phases as the disease progresses.

From a public health standpoint, animals with either form of the disease are dangerous; both may have active virus in their saliva and transmit it by biting. Although it might seem that the animal with furious rabies would be the bigger threat, at least it is easily recognized as dangerous. Animals with dumb rabies, however, are often in need of assistance and become objects of pity. But they still have the strength to bite. Indeed, a woman who went to the aid of a Silver-haired Bat she noticed flopping about on the ground was one of the first whose death was attributed to rabies spread by bats in the United States. The woman did not worry about the bite she received from the bat, and her Good Samaritan instincts were her undoing.

People should avoid any animal they observe behaving abnormally. This includes animals lying sick and apparently helpless, those becoming unnaturally aggressive or tame, or even animals wandering about at an unusual time of day. Unprovoked bites, especially those delivered by an animal behaving strangely, should receive immediate medical attention, and the animal should be reported to the health authorities. Never handle the remains of animals you find dead.

Because all mammals, including man, are susceptible to rabies, we should be extremely cautious in situations that could expose us to this frightening disease.

However, with the exception of vampire bats, the incidence of rabies in most populations of bats is too low to warrant invoking the threat of this disease as justification for anti-bat programmes. It is sobering to realize, for example, that foxes, cows, and skunks (in that order) comprise most of the rabies cases reported annually in eastern Canada. Bats rank very low on this list, at about the same level as pigs.

HISTOPLASMOSIS

Often called 'the curse of the mummy's tomb,' histoplasmosis is a fungus disease of the lungs known from many parts of the world. The symptoms are frequently similar to those of tuberculosis, although in less advanced stages they may be mistaken for those of a slight chest cold. The spores of the fungus are often found in the droppings of bats and birds, so that people visiting or working in bat or bird roosts, particularly under dry conditions, are frequently exposed to the disease. In Egypt, populations of several species of bats living in old tombs are probably responsible for accumulations of histoplasmosis spores there. Since the disease occasionally is debilitating and even fatal, it is easy to see how the symptoms came to be equated with a legendary curse.

In temperate regions, histoplasmosis spores are more often associated with birds, usually pigeons or chickens. In more than 10 years of working in the nursery colonies of Little and Big Brown Bats in Ontario, I was never exposed to histoplasmosis. Caves or mines used for hibernation pose no threat of histoplasmosis because hibernating bats do not produce droppings. With my no-histoplasmosis record from Ontario bat colonies, I went to Puerto Rico to work in bat caves there for two weeks, and came back showing a strong positive skin test for this disease. Like most people who are exposed to the fungus, I showed no symptoms. For those less fortunate, the disease can be treated, and there appears to be no way to avoid being exposed to the fungus in some situations. Histoplasmosis remains an occupational hazard for chicken farmers, pigeon fanciers, cave explorers, and people who study bats.

Keeping Bats Out

The caller was asking about bats, and I assumed it was yet another request for information on how to evict them from a building. Indeed, anyone monitoring my incoming telephone calls would suspect the only contact Canadians have with bats is the battle to keep them out of their homes. But this caller had a different problem. He was troubled by bats that were stowing away on a ship leaving the New Brunswick port of Dalhousie for Europe. I immediately suspected another case of migrating bats turning up on ships at sea. I was wrong

For several years the M.V. *Avon Forest* had been taking forest products from Dalhousie to Rotterdam and Southampton, returning to Halifax with automobiles picked up in Southampton and Le Havre. It seemed that 'everyone' connected with the operation knew about the bats travelling out from Dalhousie. There was even a story about a sailor who kept a bat as a pet. In May, 1980, the ship arrived in Southampton just before the Spring Bank Holiday weekend. To minimize the overtime bills, the ship's owners elected to leave most of the unloading until after the holiday. The decision was not well received by the dock-workers, who decided to get some measure of revenge for their lost pay. After some effort, they managed to find a bat on board the *Avon Forest*, and promptly reported it to the port health authorities.

Britain is one of the few countries in the world that is free of rabies, a situation carefully protected by stringent quarantine laws. Stowaway bats from North America, where rabies in bats is not unknown, were not welcome. The port authorities ordered the closing and fumigation of the vessel, and the resulting costs prompted the shipping company to find a way to avoid a recurrence.

On the telephone, it sounded relatively simple: find out how the bats board the ship and stop them. The enormousness of the challenge did not become apparent until a colleague and I saw the vessel and its surroundings. The ship was large, and it sat alongside an even larger warehouse, which sat in an extensive wharf area offering bats countless roosting opportunities. A survey of the area revealed a small resident population of Little Brown Bats that fed around the wharf, and probably used the ship as a night roost. Other bats might arrive on the wharf in bundles of lumber brought from a local mill as cargo or dunnage. Because the *Avon Forest* took about five days to load, there was plenty of time for bats to get used to using the hold as a night roost.

After some deliberation ruled out chemical and acoustic deterrents as impractical, we decided the only way to solve the problem was to search the ship after it left Dalhousie. We thought it would be easy to locate active bats by touring the hold with a bat detector. Surveys of the ship could be conducted often enough during the passage to ensure that no bat slipped through the screening. Once they were detected, it would be relatively easy to take steps to make sure the bats did not disembark on the other side, or come to the attention of the port authorities. This example indicates that there is rarely an easy way to solve problems with bats.

The *Avon Forest* example is not typical, of course, but it shows that problems involving bats can usually be resolved with a little patience, knowledge, and ingenuity. Bats most often come in contact with people around or inside buildings, especially those used as roosts. It follows, then, that the most effective way to minimize contact with bats is to keep them out of buildings.

I was surprised to learn that most human-bat confrontations in eastern Canada occur not in June and July when species may form nursery colonies in buildings, but in August: the peak period for requests for bat control and submission of bats for rabies testing coincides with swarming and mating season. Some bat-human contacts in late June and early July do involve nurseries, though, and most confrontations in July result from young bats learning how to fly. On their first few flights, these bats may overextend themselves and then are forced to pass the day in exposed, unsuitable roosts.

In some cases there is no reason why bats should not roost in a building, especially if no one else lives there. Unlike rodents, bats do not gnaw and cause structural damage. But, in a situation where it is important to evict them from a building, there are three possible steps: sealing them out of the building, making the conditions in the building unsuitable for them, and using poisons to eliminate them.

The only effective way to resolve a bat problem in a building is to determine how they get in and take steps to stop them. If a crack is wider than five millimetres, a bat can use it as an entrance and exit hole. But these small holes are easy to locate if large numbers of bats are using them; the accumulations of droppings on the walls, windows, or ground below provide good clues. Furthermore, regularly used openings in light-coloured walls are easily recognized by the stains around the edges. Entrance and exit holes are commonly located near the interfaces of a wall and roof, although loose chimney flashings and unscreened louvres also provide access. Bats may enter through gaps in the facerboards along the eaves or near dormer windows. And when the walls are made of stone, the bats often enter through gaps between the irregular face of the stone and the eaves.

If you do not know how the bats get in and out, let them show you. When the population is large, you will quickly be able to pinpoint their exit routes by waiting outside and watching their evening exodus. When the numbers of bats are small, though, you may not be able to locate their holes accurately. In this case, you will have little choice but to go over the building with a fine-toothed comb, blocking any hole that is larger than five millimetres, particularly those with accumulations of bat droppings near them.

Because bats do not chew or gnaw their way in, any number of light building materials will be enough to keep them out. Window screen, especially when bats enter holes required for building ventilation, caulking cement, lath, and moulding are effective blocking materials. If you are lucky, the resident bats will be using one or two conspicuous holes that are easily blocked. Sometimes when you close off one hole, the bats will show you a few more. In other cases, particularly with older houses, there may be numerous entrances for the bats and their eviction can be a project that lasts months or even years. When the bats enter under shingles or other loose roofing material, it may require major construction to keep them out. In these situations, however, the condition of the roof probably demands attention for other reasons.

You should never try to seal a bat colony inside a building. This just substitutes one problem for another, as the bats will now turn up all over the house as they try to find a way out. Whenever I set out to bat-proof a building, I try to spend several evenings locating the exit holes and counting the bats to get a picture of the size of the population. Then, on a suitable evening, I count the bats as they leave to feed, wait another 15 or 20 minutes, and then seal up the exits. This approach minimizes the number of bats you trap inside, as long as you do not do the operation when there are babies in the colony. Sealing baby bats in the colony often sends their mothers into a frenzy, prompting them to work extra hard to get back in, and perhaps undoing some of your control efforts. If the mothers are unsuccessful, it can also lead to an accumulation of dead infants. Neither situation is pleasant or necessary. Administer your control operations in early spring before the young are born, or wait until later in the summer when the colony is dispersing and the young are able to fly.

When bats are living in an attic, and blocking their entrance is not feasible, it is possible to make them feel unwelcome by changing the conditions in their attic. The best way to achieve this is to install lights and leave them on 24 hours a day throughout the 'bat season.' In one Ontario study, lighted colonies decreased in size by up to 96 per cent, while untreated groups grew by up to 97 per cent over the same period. Incandescent 100-watt bulbs appear to be the most effective, provided they are positioned so they do not cast dark shadows. Care must also be taken that the arrangement of the lights and wiring does not create a fire hazard. Of course, control by lighting is much less effective when the bats can retreat into spaces between the roof boards and to other dark nooks and crannies.

Chemical deterrents are also available to repel bats, but their use is often ineffective and may be troublesome for people. Some of these chemicals, including Roost-no-more[R] and Tanglefoot[R], are sticky substances which can be smeared around the bats' entrance areas. But if you are going to the trouble of finding the entrances, you might as well seal them off permanently. Bat traffic will eventually wear away the chemical deterrent, but a piece of screen or wood will keep them out for good. However, an application of some of these sticky materials to the material you have used as a sealant may discourage the bats from searching too hard for a way around the blockage.

Some people say that moth balls will drive bats from an attic roost. I have found that by the time you add enough moth balls to force the bats out, you have made the rooms below uninhabitable for people.

Other chemicals are designed to kill bats, but they usually do not eliminate entire colonies and may be dangerous to people and their pets. It is important to remember that it is not possible to put out poisoned baits for bats as you can for mice or rats. Bats catch live insects, and trays of poisoned seeds or other food have no particular attraction for them, although tainted fruit might provide a means of controlling frugivorous species. As a result, exterminators asked to apply poison to a colony of bats usually use a powder that includes some measure of DDT. The bats may absorb the DDT and other poisons through their skin, or ingest it while grooming. Adding poisons to a colony may reduce its numbers, but often the bats continue to use the treated site. Most of the poisons used against bats do not have an immediately immobilizing effect. The bats may become ill almost immediately, but even when exposed to high dosages, they usually live more than 24 hours. During this time, ailing bats frequently leave the treated colony, and the chances of their coming in contact with people or their pets are increased. None of the chemicals I have tested for their effectiveness at killing bats works quickly, including DDT, zinc phosphide, and fenthion, an organophosphorous insecticide.

If you are thinking of using poisons to rid yourself of bats, remember that you are both mammals, and what kills one can also kill another, often after considerable suffering. There are two other drawbacks to control methods not involving the blocking of the entrances used by the bats: the relatively long lives of the animals, and their persistent drive to find good roosts. Unless the attic they are using is permanently sealed to them, they will return when the owner forgets to leave on the lights, or when the toxin loses its effect, perhaps some years after it was applied.

Bats hibernating in buildings pose a special problem. In North America, Big Brown Bats frequently appear in the living quarters of people after arousing from hibernation and setting out in search of water. The homeowner often becomes aware of the bats only after he/she finds them in a sink or toilet bowl where they have drowned trying to get a drink of water. Controlling these bats is very difficult. The small numbers involved and the unpredictability of their comings and goings make it virtually impossible to find out how they get into the building. Furthermore, because it is difficult to know just where in the building they are hibernating, it is often not feasible to use control techniques involving lights, chemical repellants, or poison.

However, there are steps you can take to minimize contact with hibernating bats. Make sure that the entrance to the attic is well sealed; if there is a door, be certain it fits snugly, especially along the floor. If you store clothing in attic closets, make sure these, too, have tightly fitting doors. And since bats can move from the attic to the basement in the walls, be certain the door to the cellar fits tightly and is kept shut.

If you encounter an active bat in your house, be tolerant. Remember that the Chinese consider bats to be symbols of good luck and long life. Try to shut the bat into a room in which you have opened a window and removed the screen. This will give the bat a chance to leave on its own. When this is impossible, give the bat an opportunity to land somewhere and then try to catch it, perhaps in a fisherman's landing net, or by throwing a towel over it. If it has been quietly approached, it is occasionally possible to capture a roosting bat in a gloved hand; it is always wise to minimize your chances of being bitten.

Many people asking me about keeping bats out wonder why sound is not used to control them. It seems logical to presume that if you presented bats with certain sounds, distress or warning signals, or noises which might jam their echolocation, you could drive them away. But all of the sounds we have tested on Little Brown Bats either have been ignored or have attracted more bats. There does not seem to be a 'distress' call that sends bats away, and efforts to move them or jam their echolocation systems by producing high-intensity ultrasonic sound have been unsuccessful. Put simply, bats are accustomed to high-intensity sounds. Machines marketed as 'ultrasonic rodent repellers' have also proved unsuccessful with bats. In one test we ran in the laboratory, instead of being driven away, the bats preferred to roost on the machines.

Some biologists believe that bats should not be evicted from buildings under any circumstances. Of course, this is not an attitude shared by most homeowners, and control methods are obviously necessary under some circumstances. If you feel you must purge your attic of its bats, remember it is more agreeable to exclude them than to kill them by poisons or by sealing them inside.

Nowhere in the world do bats enjoy more legal protection than in Britain. There the steps I have suggested for excluding bats from buildings may be taken only after consultation with the Nature Conservancy Council. The same council must approve any efforts to evict bats, and under no circumstances does the British Wildlife and Countryside Act 1981 permit the intentional killing, injury, or capture of bats. I heartily endorse protection that precludes killing bats under the guise of 'control,' but some homeowners may be less enthusiastic about the bureaucracy of exclusion associated with this level of protection. There is no doubt that the banning of chemical control of bats is an intelligent step overdue elsewhere in the world if for no other reason than that it does not work.

If you have tried every humane method and your 150-year-old house still has more bat entrances than shingles, there may be another alternative. A musical friend from Zimbabwe has noticed to his dismay that some bats are attracted to the skirl of his bagpipes. A few members of his enthusiastic winged audience have even landed on him, and tried to crawl under his clothing, including up his kilt. But if you employ this traditional method of pest control and have someone lure them away, remember to pay the piper.

Conservation

For too long bats have suffered from a 'bad press.' In much of the world they are treated as harbingers of evil, beasts that deliberately tangle themselves in long hair and consort with supernatural beings such as werewolves and vampires. They are accused of stealing bacon and of being blind. True, bats are considered symbols of fertility and long life in some societies, but in the West, at least, their bad image persists. The reputation is understandable, in part: their roosting haunts are often dark and sinister, and because their nocturnal habits coincide with the activities of many human villains, they suffer from guilt by association. But bats do not try to tangle themselves in people's hair, nor do they steal bacon, nor are they blind.

So why the fear and misunderstanding? Certainly, widespread ignorance is the main reason. Only three species of bats are blood-eaters, yet all bats are feared as if they were vampires. The situation is compounded by the economic effects of some bats. There is little love for fruit-eating bats that prefer cash crops for their food, and there is not much compassion for vampire bats that spread rabies to livestock. Apart from this, nobody has found a way to make money from bats. The lack of sympathy for some bats has damaged the image of bats everywhere. And, as a result, populations of species in many parts of the world are in decline.

By 1978, two North American species, Grey and Indiana Bats, were listed as officially endangered because of sharp declines in their populations. Grey Bats band together to make inhospitable roosts suitable as nurseries, and large populations are essential to their survival. It is not unreasonable, then, to include Grey Bats on the Endangered List, even though their populations may reach tens of thousands in some caves.

We do not know whether other bats should be added to the Endangered List, nor are we entirely certain that placement on the list carries any particular advantages. Field studies of Indiana Bats in northern New York State suggest that legal protection does relatively little to assist the bats, as conservation officers rarely take threats to the bats seriously and people continue to visit the caves where they hibernate. Despite concerted efforts to protect Grey Bats in other areas, vandals still use fire and firearms to attack them in their roosts. Perhaps we can do more to protect bats by leaving them alone. However, successful protection must encompass all aspects of their life history, and until we know more about them, our management plans will be weak, even for species already on the List. But, in the meantime, official endangered status provides some measure of protection from massive environmental changes, such as hydroelectric developments, that could fatally affect local populations by destroying their preferred habitat.

Several factors can be identified as important threats to the survival of all bats, but it is often difficult to identify the most important element for any given species. Some combination of pesticides, disturbance, and exploitation probably constitutes the major menace in most cases. However, in tropical areas where rain forests are being destroyed, and in many temperate regions where development continues to eat up land at an alarming rate, habitat destruction is probably the principal threat to bats and many other organisms.

Bats are exposed directly to pesticides when poisons are used to control them, and indirectly through their food. In most cases, only bats living in buildings are subject to direct exposure, but any insectivorous species feeding where spraying occurs risks the ingestion of contaminated food. Healthy bats with plentiful food supplies are usually able to tolerate considerable amounts of pesticides. But, as we have seen, many bats, particularly the young, are prone to debilitating or fatal concentrations of poisons that build up when fat reserves are depleted. There is no easy, practical solution to this problem, however, for in many parts of the Third World pesticides are the main line of defence against diseases spread by insects.

While there may be little we can do to stop the world-wide poisoning of bats by pesticides, we should be able to halt disturbances to them in their roosts. Secure roosts are vital to their survival, and even the smallest disturbance will often force them to abandon their sites. Arousal from torpor compromises the exact energy-saving budgets of hibernating bats, and these disturbances can lead to the deaths of huge numbers of bats. As a result, efforts should be made to strictly curtail recreational cave exploration in places where it disturbs bats at critical times. A small amount of consideration by cave explorers could make a large contribution to the survival of bats.

But people who study bats, particularly in their hibernation sites and nursery colonies, have also had a serious impact on populations of bats. In fact, the problem was grave enough to prompt the U.S. Fish and Wildlife Service to place a moratorium on bat banding to minimize the disturbance the animals suffered. Responsible bat biologists strictly limit the agitation associated with their activities by banding only when absolutely necessary, and then away from roosts; bats captured on their way to or from feeding areas rarely abandon local roosts, and in studies of mother-young interactions, in nursery colonies, examination and banding of the young can be done after the mothers have left. Refraining from disturbing bats does not preclude further research

Exploitation of large bat populations for scientific or educational purposes has also contributed to the decline in numbers of some species. Many European and North American sites that once harboured huge populations of bats now have none. In some cases, the disappearance can be directly related to studies in which impressively large samples produced more convincing results. In others, the bats were collected by biological supply companies and sold to educational institutions to add a new dimension to science courses.

Of course, bats' low reproductive rates are not compatible with this type of exploitation. This does not mean that students cannot enjoy the thrill of watching bats; they can grab a ringside seat simply by visiting local streams and ponds, or even by waiting under a street-light where the animals come to feed. This type of outing, admittedly outside of regular school hours, could be part of a programme involving the class in the construction of a bat detector, a microphone sensitive to high-frequency sounds. Not only would such a programme blend together several facets of science, from physics to biology, it also would not disturb the bats.

The existing knowledge about populations of bats does not permit accurate identification of species in danger of extinction. When there are declines in the numbers of a particular species, it is usually impossible to determine the principal factor responsible. However, we should take immediate steps to stop the use of poisons against bats, if for no other reason than that they are largely ineffective. Bats should be protected from disturbance, too. Any of these steps involves education of the public, cave explorers, and people who study bats. In parts of Europe, the renovation of old buildings harbouring bats is controlled to protect the animals wherever possible. Excluding bats from their roosts as a means of control may not benefit the populations, but it at least provides the animals with a chance to re-establish themselves. Control by killing offers no such opportunity. In Europe and Australia, biologists are experimenting with bat houses. Similar in basic design to a bird house, they have been exploited by some species of bats and may prove important in bat conservation.

Convincing people that bats deserve protection means we must first help them to overcome widespread fear and ignorance. Many bats consume vast quantities of insects each night, and studies of their feeding behaviour may one day lead to their use in controlling populations of harmful insects. But even if we are never able to manipulate them to help us directly, bats are worthy of protection. They represent a classic example of how animals operate over a wide range of circumstances. Their lengthy list of specializations, including exploitation of new opportunities, echolocation, mating strategies, and capacity to survive environmental extremes, make them as fascinating as any creatures on earth.

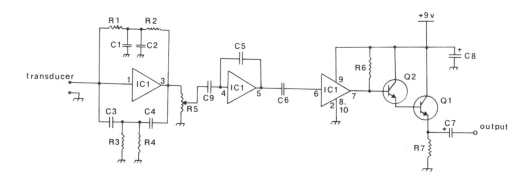

This circuit diagram shows the arrangement of components of a simple bat detector, suitable for detecting sounds at 40 kHz. Available commercially as detectors for finding leaks in gas lines, these are easily adapted to the study of the echolocation calls of some bats. To construct one, the following components are essential: resistors – R1, R2, R3, R4 all 3.9 K ¼ Watt 5%; R5 2.5 K linear potentiometer; R6 27 K ¼ Watt; R7 10 K ¼ Watt; capacitors – C1, C2, C3, C4 all 0.001 μfd, 50 v ceramic, C5 100 pf 50 v ceramic, C6 0.05 μfd, 50 v ceramic, C7 10 μfd 25 v tantalum, C8 100 μfd 16 v tantalum, C9 0.05 μfd, 50 v ceramic; Q1 – 2N3567; ICI CA 3035; Q2 2N3903; miscellaneous battery clip for 9-volt source, phono jack and a Massa TR896 transducer. With the exception of the transducer, all of these components are commonly available at electronic supply shops.

Common and Scientific Names

ARRANGED BY FAMILY

COMMON NAME	SCIENTIFIC NAME
FLYING FOXES, OLD WORLD FRUIT BATS	PTEROPODIDAE
Bare-backed Fruit-eaters	*Dobsonia* spp.
Black Flying Fox	*Pteropus alecto*
Buettikofer's Fruit Bat	*Epomops buettikoferi*
Dawn Bat	*Eonycteris spelaea*
Dog-faced Bats	*Rousettus* spp.
Egyptian Fruit Bat	*Rousettus aegyptiacus*
Gambian Epauletted Fruit Bat	*Epomophorus gambianus*
Grey-headed Flying Fox	*Pteropus poliocephalus*
Hammer-headed Bat	*Hypsignathus monstrosus*
Harpy Fruit Bat	*Harpyionycteris whiteheadi*
Lesser Tube-nosed Bat	*Paranyctimene raptor*
Sphinx's Short-faced Fruit Bat	*Cynopterus sphinx*
Straw-coloured Fruit Bat	*Eidolon helvum*
Streseman's Dog-faced Bat	*Rousettus stresemani*
Taylor's Flying Fox	*Pteropus tablasi*
Woermann's Long-tongued Fruit Bat	*Megaloglossus woermanni*
SHEATH-TAILED BATS	EMBALLONURIDAE
Greater White-lined Bat	*Saccopteryx bilineata*
Lesser White-lined Bat	*Saccopteryx leptura*
Mauritian Tomb Bat	*Taphozous mauritianus*
BUTTERFLY BATS	CRASEONYCTERIDAE
Butterfly Bat	*Craseonycteris thonglongyai*
BULLDOG BATS	NOCTILIONIDAE
Bulldog Bats	*Noctilio* spp.
Lesser Bulldog Bat	*Noctilio albiventris*
Mexican Bulldog Bat	*Noctilio leporinus*

SLIT-FACED BATS
Egyptian Slit-faced Bat
Large Slit-faced Bat
Wood's Slit-faced Bat

NYCTERIDAE
Nycteris thebaica
Nycteris grandis
Nycteris woodi

FALSE VAMPIRE BATS
Ghost Bat
Heart-nosed Bat
Indian False Vampire Bat

MEGADERMATIDAE
Macroderma gigas
Cardioderma cor
Megaderma lyra

HORSESHOE BATS
Blasius' Horseshoe Bat
Bushveld Bat
Dent's Horseshoe Bat
Eastern Horseshoe Bat
Greater Horseshoe Bat
Hildebrandt's Horseshoe Bat
Lander's Horseshoe Bat

RHINOLOPHIDAE
Rhinolophus blasii
Rhinolophus simulator
Rhinolophus denti
Rhinolophus megaphyllus
Rhinolophus ferrumequinum
Rhinolophus hildebrandti
Rhinolophus landeri

OLD WORLD LEAF-NOSED BATS
African Trident-nosed Bat
Giant Leaf-nosed Bat
Great Round-leaf-nosed Bat
Noack's Leaf-nosed Bat
Triple Leaf-nosed Bat

HIPPOSIDERIDAE
Cloeotis percivali
Hipposideros commersoni
Hipposideros armiger
Hipposideros ruber
Triaenops persicus

MOUSTACHE BATS
Blainville's Leaf-chinned Bat
Naked-backed Moustache Bats
Parnell's Moustache Bat

MORMOOPIDAE
Mormoops blainvilli
Pteronotus spp.
Pteronotus parnellii

NEW WORLD LEAF-NOSED BATS
Big Fruit-eating Bat
California Leaf-nosed Bat
Common Vampire Bat
Fringe-lipped Bat
Geoffroy's Tailless Bat
Hairy-legged Vampire Bat
Honduran White Bat
Linnaeus' False Vampire Bat
Peters' False Vampire Bat

PHYLLOSTOMATIDAE
Artibeus lituratus
Macrotus californicus
Desmodus rotundus
Trachops cirrhosus
Anoura geoffroyi
Diphylla ecaudata
Ectophylla alba
Vampyrum spectrum
Chrotopterus auritus

Sanborn's Long-tongued Bat   *Leptonycteris sanborni*
Short-tailed Fruit Bat   *Carollia perspicillata*
Spear-nosed Bat   *Phyllostomus hastatus*
White-lined Tailless Bat   *Vampyrodes caraccioli*
White-winged Vampire Bat   *Diaemus voungi*
Wrinkle-faced Bat   *Centurio cenex*

NEW WORLD DISC-WINGED BATS   THYROPTERIDAE
Honduran Disc-winged Bat   *Thyroptera discifera*
Spix's Disc-winged Bat   *Thyroptera tricolor*

OLD WORLD DISC-WINGED BATS   MYZOPODIDAE
Old World Disc-winged Bat   *Myzopoda aurita*

PLAIN-NOSED BATS   VESPERTILIONIDAE
Australian Little Brown Bat   *Eptesicus pumilus*
Big Brown Bat   *Eptesicus fuscus*
Club-footed Bats   *Tylonycteris* spp.
Common Bent-winged Bat   *Miniopterus schreibersi*
Gould's Long-eared Bat   *Nyctophilus gouldi*
Grey Bat   *Myotis grisescens*
Hoary Bat   *Lasiurus cinereus*
Indiana Bat   *Myotis sodalis*
Lesser Yellow House Bat   *Scotophilus leucogaster*
Little Brown Bat   *Myotis lucifugus*
Little Pipistrelle   *Pipistrellus tenuis*
Long-eared Bat   *Plecotus auritus*
Long-legged Myotis   *Myotis volans*
Mexican Fishing Bat   *Myotis vivesi*
Mexican Long-eared Bat   *Myotis auriculus*
Mouse-eared Bats   *Myotis* spp.
Noctule   *Nyctalus noctula*
Northern Long-eared Bat   *Myotis septentrionalis*
Pallid Bat   *Antrozous pallidus*
Pipistrelle   *Pipistrellus pipistrellus*
Red Bat   *Lasiurus borealis*
Serotine   *Eptesicus serotinus*
Silver-haired Bat   *Lasionycteris noctivagans*
Small-footed Bat   *Myotis leibii*
Spotted Bat   *Euderma maculatum*
Western Big-eared Bat   *Plecotus townsendii*

Woolly Bats	*Kerivoula* spp.
Yellow Bat	*Lasiurus ega*
Yuma Myotis	*Myotis yumanensis*

FREE-TAILED BATS	MOLOSSIDAE
Ansorg's Free-tailed Bat	*Tadarida ansorgei*
Big Free-tailed Bat	*Tadarida macrotus*
Duke of Abruzzi's Free-tailed Bat	*Tadarida aloysiisabaudiae*
Greater House Bat	*Molossus ater*
Greater Mastiff Bat	*Eumops perotis*
Long-crested Free-tailed Bat	*Tadarida chapini*
Martienssen's Free-tailed Bat	*Otomops martiensseni*
Mexican Free-tailed Bat	*Tadarida brasiliensis*
Naked Bat	*Cheiromeles torquatus*
Pallas' Mastiff Bat	*Molossus molossus*
Peters' Flat-headed Bat	*Platymops setiger*
Roberts' Flat-headed Bat	*Sauromys petrophilus*
South American Flat-headed Bat	*Neoplatymops mattogrossensis*
Thomas' Mastiff Bat	*Eumops trumbulli*

Sources of More Information

Allen, G.M. 1939. *Bats*. Harvard University Press, Cambridge

Altenbach, J.S. 1979. *Locomotor morphology of the vampire bat, 'Desmodus rotundus.'* Special Publication no. 6, American Society of Mammalogists

Baer, G.M. (editor). 1975. *The natural history of rabies*. Academic Press, New York

Baker, R.J., J.K. Jones Jr, and D.C. Carter (editors). 1976. *Biology of bats of the New World family Phyllostomatidae*. Part 1. Special Publication, The Museum, Texas Tech University, Lubbock, Texas

– 1977. *Biology of bats of the New World family Phyllostomatidae*, Part 2. Special Publication, The Museum, Texas Tech University, Lubbock, Texas

– 1979. *Biology of bats of the New World Family Phyllostomatidae*. Part 3. Special Publication, The Museum, Texas Tech University, Lubbock, Texas

Barbour, R.W. and W.H. Davis. 1969. *Bats of America*. University of Kentucky Press, Lexington, Kentucky

Brosset, A. 1966. *La biologie des chiroptères*. Masson et Cie, Paris

Busnel, R-G., and J.F. Fish (editors). 1980. *Animal sonar systems*. Plenum Press, New York

Griffin, D.R. 1958. *Listening in the dark*. Yale University Press, New Haven

Gustafson, A.W. and B.J. Weir (editors). 1979. 'Comparative aspects of reproduction in Chiroptera.' *Journal of Reproduction and Fertility*. Symposium Report no. 14

Hall, L.S. and G.C. Richards. 1979. *Bats of eastern Australia*. Queensland Museum Booklet no. 12, Brisbane

Humphrey, S.R. and J.B. Cope. 1976. *Population ecology of the little brown bat, 'Myotis lucifugus,' in Indiana and north-central Kentucky*. Special Publication no. 4, American Society of Mammalogists

Husson, A.M. 1962. *The bats of Suriname*. Zoologische Verhandelingen no. 58, Leiden

Irvine, A.D., J.E. Cooper, and S.R. Hedges. 1972. 'Possible health hazards associated with the collection and handling of post-mortem zoological material.' *Mammal Review*, 2:43–54.

Kingdon, J. 1974. *East Afrian mammals, an atlas of evolution in Africa.* Volume 2A. Academic Press, London

Kunz, T.H. (editor). 1982. *Ecology of bats.* Plenum Publishing Corp. New York

Leen, N. and A. Novick. 1969. *The world of bats.* Holt Rinehart and Winston, New York

Marshall, A.G. 1981. *The ecology of ectoparasitic insects.* Academic Press, London

Miller, G.S. 1907. *Families and genera of bats.* United States National Museum Bulletin 57, Washington

Pennycuick, D.J. 1972. *Animal flight.* Arnold Studies in Biology no. 33

Peterson, R. 1964. *Silently by night.* Longmans, London

Roeder, K.D. 1967. *Nerve cells and insect behavior,* revised edition. Harvard University Press, Cambridge

Rosevear, D.R. 1965. *The bats of west Africa.* Trustees of the British Museum (Natural History), London

Schmidt, U. 1978. *Vampirfledermäuse.* Die Neue Brehmn-Bücherei, A. Ziemsen Verlag, Wittenberg Lutherstadt

Silva Taboada, G. 1979. *Los murcielagos de Cuba.* Editorial Academia, Havana

Slaughter, B.H. and D.W. Walton (editors). 1970. *About bats, a chiropteran symposium.* Southern Methodist University Press, Dallas

Strickler, T.L. 1978. *Functional osteology and myology of the shoulder girdle in the Chiroptera.* S. Karger, Basel

Turner, D.C. 1975. *The vampire bat, a field study in behavior and ecology.* Johns Hopkins University Press, Baltimore

Villa-R, B. 1966. *Los murcielagos de Mexico.* Universidad Nacional Autonoma de Mexico, Instituto de Biologia, Mexico

Wallin, L. 1969. 'The Japanese bat fauna.' *Zoologiska Bidrag Fran Uppsala,* 37:223–440

Wilson, D.E. and A.L. Gardner (editors). 1980. *Proceedings of the fifth international bat research conference.* Texas Tech University Press, Lubbock

Wimsatt, W.A. (editor). 1970. *Biology of bats,* volume 1, Academic Press, New York

– 1970. *Biology of bats,* volume 2. Academic Press, New York

– 1977. *Biology of bats,* volume 3. Academic Press, New York

Yalden, D.W. and P.A. Morris. 1975. *The lives of bats.* Demeter Press, Quadrangle, New York Times Book Co.

SOME SPECIALIST READINGS

Bradbury, J.W., D. Morrison, E. Stashko, and R. Heithaus. 1979. 'Radio-tracking methods for bats.' *Bat Research News,* 30:9–17

Buchler, E.R. 1976. 'A chemiluminescent tag for tracking bats and other small, nocturnal animals.' *Journal of Mammalogy,* 57:173–6

Greenhall, A.M. and J.L. Paradiso. 1968. *Bats and bat banding*. United States Department of the Interior, Fish and Wildlife Service, Bureau of Sport Fisheries and Wildlife, Washington

Nagorsen, D.W. and R.L. Peterson. 1980. *Mammal collectors' manual*. Life Sciences Miscellaneous Publications, Royal Ontario Museum, Toronto

Simmons, J.A., M.B. Fenton, W.R. Ferguson, M. Jutting, and J. Palin. 1979. *Apparatus for research on animal ultrasonic signals*. Life Sciences Miscellaneous Publications, Royal Ontario Museum, Toronto

Tidemann, C.R. and D.P. Woodside. 1978. 'A collapsible bat-trap and a comparison of results obtained with the trap and with mist nets.' *Australian Wildlife Research*, 5:355-62

Tuttle, M.D. 1974. 'An improved trap for bats.' *Journal of Mammalogy*, 55:475-7

SOME BOOKS FOR CHILDREN

Derennes, C. 1924. *The life of the bat*. Hooper and Brothers, New York

Freeman, D. 1970. *Hattie, the backstage bat*. The Viking Press, New York

Kohn, B. 1979. *The bat book*. Dandelion Books

Leen, N. 1976. *The bat*.

Mohr, C.E. 1976. *The world of the bat*. J.B. Lippincott Co., New York

Pye, D. 1968. *Bats*. The Bodley Head, London

Ripper, C.L. 1954. *Bats*. William Morrow and Co., New York

Ungerer, T. 1961. *Rufus*. Harper, New York

Index

Italic page numbers refer to illustrations.

This book

was designed by

ANTJE LINGNER

of University of

Toronto

Press